Astronomers' Observing Guides

For other titles published in this series, go to
www.springer.com/series/5338

Other Titles in This Series

Star Clusters and How to Observe Them
Mark Allison

Saturn and How to Observe it
Julius Benton

Nebulae and How to Observe Them
Steven Coe

The Moon and How to Observe It
Peter Grego

Supernovae and How to Observe Them
Martin Mobberley

Double & Multiple Stars and How to Observe Them
James Mullaney

Galaxies and How to Observe Them
Wolfgang Steinicke and Richard Jakiel

Total Solar Eclipses and How to Observe Them
Martin Mobberley

Cataclysmic Cosmic Events and How to Observe Them
Martin Mobberley

Martin Mobberley

The Caldwell Objects and How to Observe Them

with 183 Illustrations

Springer

Martin Mobberley
Denmara
Cross Green
Cockfield
Bury St. Edmunds
Suffolk. IP30 0LQ
United Kingdom
martin.mobberley@btinternet.com

Series Editor
Dr. Mike Inglis, BSc, MSc, Ph.D.
Fellow of the Royal Astronomical Society
Suffolk County Community College
New York, USA
inglism@sunysuffolk.edu

ISBN 978-1-4419-0325-9 e-ISBN 978-1-4419-0326-6
DOI 10.1007/978-1-4419-0326-6
Springer Dordrecht Heidelberg London New York

Library of Congress Control Number: PCN applied for

Printed on acid-free paper

Springer is part of Springer Science+Business Media (www.springer.com)

Preface

When Mike Inglis, who consults for Springer, first asked me to write a Caldwell book in their "and How to Observe Them" series I admit I did need some time to think the suggestion over. I am a fan of Patrick's Caldwell catalog as are most of the amateur astronomers I know, but could a new book be justified when the massive, comprehensive, and wrist-spraining Caldwell tome by Stephen O'Meara covers the subject fully anyway? That other book was researched and written over a 5-year period in the pre-9/11 world from 1996 to 2001, and its 484 pages of descriptive text and background data are a joy to peruse, as are Stephen's impressive sketches. OK, the book is far too heavy for the binding and after a few trips outside half the pages in my copy fell out, but apart from that it is an excellent book and surely impossible to improve upon, or even equal, especially in a smaller format book with only half the pages available; at least, that was my initial impression when I mulled over writing another Caldwell book. However, I changed my mind, because a number of events of astronomical significance have occurred in the last 10 years.

First and foremost backyard imaging of deep sky objects, especially color imaging, has come on in leaps and bounds in the twenty-first century. In addition image processing power and the size of affordable CCD detectors have both accelerated rapidly, as have developments in commercial color printing. This has all opened the doors for a new book using mouthwatering Caldwell object images, many reproduced in color, acquired by deep sky specialists worldwide. Astronomy has experienced a huge surge in interest in recent years, mainly at the CCD imaging end of the market and this book, unlike Stephen O'Meara's tome, caters more for the deep sky imager, rather than the visual observer. Although Stephen's sketches are excellent, there are very few observers with such drawing skills, but many more with CCD cameras and digital SLRs.

Also, significant advances in cosmology have taken place in the twenty-first century. Toward the end of the twentieth century, the first hints were coming through that the redshifts of distant Type Ia supernovae were rather confusing. Suggestions were being made that the universe had actually started to accelerate its expansion in recent eons. What does this have to do with the Caldwell objects, you may say? Well, the current thinking is that, because of the resulting knowledge deduced from the redshift of distant supernovae and the background radiation from the Big Bang, the universe must be close to 13.7 billion years old. This age restriction affects our understanding of objects like, for example, globular clusters, of which there are 18 in this catalog; 10 years ago cosmologists (and authors) were listing these elderly starballs as being up to 16 billion years old, that is, older than the universe! So even in the space of 10 years our knowledge of the age of many deep sky objects has had to be revised.

There is another reason to write this book, too. I live in the same country as Patrick Moore, roughly the same distance north of London as Patrick is to the capital's south. Even at my country observatory, the skies are much brighter than they were 40 years ago, and the sort of skies enjoyed by Stephen O'Meara on Hawaii are never seen here, or at Patrick's Selsey base. Indeed, most UK nights are cloudy, some entire months are cloudy, and even when it is clear the air is frequently hazy or moisture-sodden. Writing this book from the damp and murky depths of the miasmic swamp of Suffolk, UK, gave me an excuse to battle with the elements over the course of a year and to visually track down all the objects in the Caldwell catalog from number 1 to 65 that I could see from my latitude. I also enjoyed seeking out the best images of those Caldwell objects that cannot rise from the UK, many of which I have never seen, despite a few astronomy trips south of the equator. After hauling Stephen O'Meara's book out to my telescopes, in the course of writing this book, even I started to wish I had a lighter, more concise Caldwell book to lug outside!

Finally, what really clinched this book project for me was when I asked Patrick's opinion, at his Selsey home, on February 20, 2008. He was very keen for me to write a Caldwell book and so, as a lifelong fan, that pretty much sealed it for me.

Suffolk, UK
July 2009

Martin Mobberley

Acknowledgements

I am, as always, indebted to all those who have helped me with free images and information. Without such generosity very few of the books aimed at amateur astronomers would be financially viable at all. I give my sincere thanks to the following contributors to this book: Ron Arbour, Adam Block & Flynn Haase (AOP), Tom Boles, Tom Carrico, Till Credner (AlltheSky.com), Christian Fuchs, Philipp Keller and the T1T-Team, Robert Gendler, Tom Harrison, Patrick Moore, Stewart Moore, Josef Pöpsel and Stefan Binnewies of Capella Observatory, Gordon Rogers, Ian Sharp, Daniel Verschatse, Volker Wendel and Bernd Flach-Wilken of the Spiegel team and Geof Wingham. Like so many other authors I am also indebted to NASA, ESA, those associated with the Hubble Space Telescope (The Space Telescope Science Institute – STScI) and the Digitized Sky Survey team (image copyright's held by AAO Board/PPARC/CIT/AURA) for the use of their images. Thanks also to Mike Inglis, John Watson, Patrick Moore, and Richard Tresch Fienberg for giving their views on any copyright issues surrounding existing published Caldwell data.

I also thank all at Springer New York, but especially Jenny Wolkowicki who has supervised the entire production process and worked with me on a total of five books now. Finally, I thank my father Denys for his constant support in all my observing and writing projects.

About the Author

Martin Mobberley is a well-known British amateur astronomer, who observes and images a wide variety of objects, including comets, planets, novae, supernovae, and asteroids. He has written six previous amateur astronomy and practical observing guides for Springer as well as three children's books about astronomy and space travel. In 1997, the International Astronomical Union (IAU) named asteroid number 7239 as "Mobberley" in recognition of Martin's contribution to amateur astronomy. Martin served as the president of the British Astronomical Association from 1997 to 1999, and he received the association's Goodacre award in 2000. He is a regular contributor to the British magazine *Astronomy Now* and has appeared as a guest on Sir Patrick Moore's long-running *Sky at Night* BBC TV program on numerous occasions since the late 1990s.

Contents

Chapter 1

Sir Patrick Moore, Observer Extraordinaire

Introduction

Sir Patrick Moore, (Fig. 1.1) or, if you really prefer, Sir Patrick Caldwell-Moore, the originator of the Caldwell catalog, is best known for being the most prolific author of astronomy books the world has ever known and for presenting the BBC *Sky at Night* television series since April 1957. If you stop someone in the street in the UK and ask them to name a famous astronomer, they will invariably say "Patrick Moore" instantly, with Professor Stephen Hawking coming a distinct second. I first met Patrick at a British Astronomical Association (BAA) Lunar Section meeting at the Royal Geological Society in Piccadilly, London in September 1970, when I was aged just 12. In my teenage years, I attended many meetings where Patrick was a speaker. Since the early 1980s our paths have crossed many times at BAA subcommittee and council meetings, and on the numerous occasions I have visited his home, or been thrilled to appear with him on the *Sky at Night*. So, I think, almost 40 years after first meeting him, and almost 30 years after first working with him at the BAA, I know him pretty well.

Patrick is that very rare specimen, a man of principle. If he agrees to help you with something and shakes your hand, it is an unbreakable agreement; he never goes back on his word. He is also one of the most generous people I have ever met, yet becomes highly embarrassed if anyone mentions this or the voluntary work he has done for British Astronomy and numerous charities. Many a time I have attended a public lecture by the great man and the chairman or compére has announced Patrick is waiving his fee or donating it to a charity, to thunderous applause. Patrick, like most British amateur astronomers, has no time for political correctness, and he firmly believes that Great Britain is the greatest nation on Earth. Occasionally his forthright views have produced media criticism, but whereas his critics have come and gone, Patrick soldiers on. Every year, since the early 1950s, he has churned out more and more books and articles and his monthly TV program is by far the longest running program with the same presenter in the entire world. Sheer enthusiasm for astronomy (and a bit of luck) is what thrust Patrick into the astronomy limelight during the dawn of the space race and the *Apollo* lunar landing era. Half a century later, sheer enthusiasm has kept him there.

However, even I did not realize just how much visual observing, at the telescope eyepiece, Patrick had carried out until I had spent many happy hours locked away

M. Mobberley, *The Caldwell Objects and How to Observe Them*,
DOI 10.1007/978-1-4419-0326-6_1, © Springer Science+Business Media, LLC 2009

Fig. 1.1. Patrick Moore on the set of his TV program "The Sky at Night." Photograph taken by the author in December 1999.

in his astronomy notebook library at his home in Selsey (Fig. 1.2). Patrick still has all his observing notebooks dating back to the early 1930s. They are divided up into different categories such as Solar, Lunar, Venus, Mars, Jupiter, Saturn, Variable Star, etc., and there are maybe a dozen or more books in each category, crammed to the brim with sketches and notes.

Although Patrick has mainly been a lunar and planetary observer, he has observed the deep sky in earnest, too. Frankly, I would not hesitate to suggest that he was the greatest all-around visual observer of the twentieth century. Although there are variable star observers who have made more than 100,000 magnitude estimates, and comet hunters who have discovered many of those fuzzy objects with nothing more than their dark adapted eyes and modest telescopes, I do not believe there is anyone who has made so many visual observations of such a wide variety of differing objects. His observing stamina is legendary, too. Leaf through his observing log books for Jupiter and you can find nights when the planet was at a high declination when he timed *two* transits of Jupiter's Red Spot in one night, almost 10 h apart, in subzero conditions, in a continuous observing stint when every longitude of the planet was sketched! Friends observing with him have often commented that they would all be wrapped up with jumpers, winter coats, gloves, and balaclavas, whereas Patrick would just stroll outside with his blazer on, maybe just draping a scarf around

Fig. 1.2. The noted planetary observer Damian Peach mesmerised by just one of Patrick's dozens of visual observing notebooks, spanning over 70 years, in Patrick's home "Farthings" at Selsey, UK.

his neck as a token acknowledgement of the cold. I have even witnessed him dressed thus on bone-chilling nights, in his 80s, with a streaming cold, hobbling outside with walking sticks, or in a wheelchair, while all those around, decades younger than the great man, were wrapped up like arctic explorers, constantly shivering!

Patrick's Telescopes

From the early 1930s Patrick used a trusty 3-in. refractor and then, after the war years, he acquired a fine alt-azimuth mounted 12½-in. f/6 Newtonian (Figs. 1.3 and 1.4). An equatorially mounted 8-in. f/7 Newtonian was added in the 1950s, a 5-in. Cooke refractor in the 1960s, and a 15-in. f/6 Newtonian in the 1970s. In addition to these instruments, Patrick has observed through a staggering number of other famous telescopes worldwide, such as the 24-in. Lowell refractor near Flagstaff, Arizona. So his experience is not only wide-ranging in terms of objects but also in terms of telescopes used, too.

In recent years, Patrick's 15-in. f/6 Newtonian has been his flagship instrument and, as I have used this telescope a number of times, I would like to mention why it is such a special instrument for the visual deep sky observer. In an era when production-line Schmidt-Cassegrains seem to dominate and everyone's telescope looks the same, Patrick's 15-in. is a breath of fresh air (Fig. 1.5). The octagonal

Fig. 1.3. A photograph probably taken around 1955 showing Patrick at the eyepiece of his 12.5-in. alt-azimuth mounted f/6 Newtonian, a telescope which Patrick has owned for over 60 years! Reproduced by permission of Patrick.

wooden tube features a rotating top end such that the eyepiece can be pulled around to any position for user comfort. In addition, the 7-foot-long tube, already quite heavy at the mirror end, has extra weights mounted near the mirror cell, allowing a very low-slung fork design to be used, thus keeping the 90-in. long focal length instrument's eyepiece within reach (with the help of some wooden steps) even when pointing near the zenith. For most celestial objects this low-slung design, long tube, and rotating top end enable a standing observer to be relatively comfortable, without falling foul of a plinth or having to climb many rungs of a ladder. The finder telescopes, carefully positioned near the eyepiece on the rotating top end, add to the user-friendly capability. Although the 15-in. instrument does not have a modern "Go To" facility, the large aperture, comfortable viewing position, and big declination circle make locating objects easy. To facilitate a faster cool down after a warm day, numerous parts of the octagonal tube can be removed, so the closed tube more resembles an open tube and air can circulate freely, eliminating tube currents. Unusually, the instrument features two right ascension worm and wheel units, one for fast slewing and one for sidereal tracking.

Fig. 1.4. The author with Patrick's 12.5-in. f/6 Newtonian in the great man's garden at Selsey, UK. Photograph taken by Damian Peach with the author's camera in March 2005, some 50 years (!) after the telescope appeared in the previous figure.

Fig. 1.5. Mars expert Richard McKim observing Mars with Patrick's 15-in. (38 cm) f/6 fork mounted Newtonian in October 2005. The telescope has a rotating top end, which makes comfortable viewing much easier.

Patrick's Enthusiasm

Many people have expressed views on Patrick or on his Caldwell catalog but, frankly, unless you get to know him over many years you simply will not have any idea what he is like "in the flesh." You may get a hint from his books, or his TV appearances, but no more than that. As someone who attended any convenient talk on astronomy Patrick gave in my area, from when I was 12 and older (when he was in his late forties and older), Patrick was just a sheer nuclear cauldron of energy and enthusiasm where astronomy was concerned. Most speakers at astronomy meetings can be rather boring, and even with a microphone it can be hard to hear them mumbling and laboring each point. Patrick never used a microphone in his prime because his voice was a human public address system! No-one ever missed a word he said or went away disappointed. He never refused any invitation to give a talk, and other astronomers were always welcome at his home. Patrick has no time for bureaucracy, red tape, or politicians because he is a man of action, not words. If you ask him to do something he does it *instantly*. There are no towering in-trays in Patrick's study because if something comes in it is soon dealt with!

One example I would like to share with the reader that springs to mind to illustrate this aspect of Patrick concerns a BAA council meeting I attended in London during the late 1980s. A BAA member had died and, although not a famous member, had quietly contributed a number of useful observations over many years. The association's journal editor had decided we needed an obituary writing, and a debate ensued around the council table as to who would be the best person to write the obituary. Various council members who knew the deceased admitted to knowing the man but invariably commented that "I just don't have the time at present to write an obituary." What was meant to be a minor point dragged on for about 10 min of waffle and hot air at the table, with various excuses being made for individuals *not* writing the obituary. Finally, the president banged his gavel on the table and said we had discussed this *ad nauseum* and, as no-one had time to write the obituary he would like to move on. At this point Patrick's hand shot up. "Yes Patrick?" said the president, curious as to what additional information Patrick could possibly add. Patrick boomed back across the table: "Mr President, while you were discussing this matter I found time to write the obituary of the gentleman in question. The editor can use this, or bin it, as she wishes," and Patrick passed his obituary up the table to the editor to much amusement. The obituary appeared, virtually unchanged, a few months later in the journal.

The Caldwell Concept

The great man certainly does not hang about. When he decides to do something it is started and finished in a very short time; entire books have been written in the space of a fortnight! So it was with the Caldwell concept. The idea came to Patrick after he had been observing various objects, including the double cluster in Perseus (Caldwell 14) and the Helix nebula in Aquarius (Caldwell 63). One is in the far north of the sky and easily visible from the UK. The other is a tricky object from UK latitudes but visible in September and October if you have a good southern horizon. But both can be enjoyed with small telescopes. Two things seem to have struck Patrick. The first was that there were objects in the night sky that were just as good as the 109 Messier objects but were being neglected, at least by the casual observer. Second, why not go beyond Messier and venture down to the deep southern sky and list the objects in a new catalog numbered in descending declination order? Patrick rattled off a list of his own 109 favorites and posted the list to *Sky & Telescope*, not knowing if they would like it or not. They loved it! It was published in the December 1995 edition (pp. 38–43) alongside an article (strangely forgotten these days) written by M. Barlow Pepin. Accompanying the list of 109 objects were long exposure black & white and color photographs of the Bubble nebula NGC 7635 in Cassiopeia (Caldwell 11), the prolific supernova producing galaxy NGC 6946 in Cepheus (Caldwell 12), the Cocoon nebula IC 5146 in Cygnus (Caldwell 19), the Helix nebula NGC 7293 in Aquarius (Caldwell 63), the Sculptor galaxy NGC 253 (Caldwell 65), and the Fornax galaxy NGC 1097 (Caldwell 67).

There were no sketches at all in that original article, just some very fine photographs taken by a few leading astrophotographers of the 1990s. Patrick's choice of favorite non-Messier objects is just that, *his* favorites and so, not surprisingly, there are some constellations that are favored and others that are not. Of course, some constellations, especially in the Milky Way, will have far more deep sky objects to offer, and some are simply bigger than others. Thus the constellations of Cassiopeia, Cygnus, and Centaurus each have six of Patrick's favorite objects, and Cepheus and Carina also have five apiece. Perhaps surprisingly, Canes Venatici has four Caldwell objects quite close together, but Virgo has only one (there are actually seven Caldwell galaxies close together near the Canes Venatici/Coma Berenices border). But, as Patrick said to me in his study in 2008, "It was just a bit of fun. I never dreamed it would be so popular."

Now, it goes without saying that none of the objects in the Caldwell catalog were discovered by Patrick. They all have other identities with NGC, IC, Collinder, or even Sharpless designations. Some naked-eye Caldwell objects have been known since antiquity, and 50 fainter Caldwell objects were, not surprisingly, first discovered by the indefatigable William Herschel, whose son John discovered a further six (mainly in the southern hemisphere) and sister Caroline a further three. Once we are in the far southern celestial hemisphere other tireless observers tend to dominate. James Dunlop discovered seventeen of the southernmost Caldwell objects, and Abbé Nicolas Louis de Lacaille bagged a dozen. The remainder can be credited to observers and photographers such as E.E. Barnard, Max Wolf, Sharpless, Collinder, Hodierna, Denning, d'Arrest, Schaeberle, Marth, von Auwers, and al-Sufi. Needless to say, as Patrick specifically excluded all the Messier objects from his Caldwell list Charles Messier could not have discovered any of them. In fact, by the 1990s the 109 objects in Messier's famous list were, perhaps, getting a bit too familiar to the experienced visual observer and, while the addictive writings of Walter Scott Houston in *Sky & Telescope* encouraged many to seek out more challenging objects, Patrick's idea of a separate list of 109 objects just seemed to catch on. Of course, with Patrick being so well known this helped enormously, as did the challenge of extending the list into the far southern hemisphere, where Messier had not searched.

Other amateurs had drawn attention to neglected southern hemisphere objects before. The comet and supernova discoverer Jack Bennett compiled his own "Bennett's catalog" of 152 objects from two lists he had compiled in 1969 and 1974 to help him avoid objects that looked like comets. Of course, this was the whole point of Messier's catalog, too. Many of Bennett's listings are in Patrick's Caldwell catalog. Bennett also included 26 Messier objects within his list. But Patrick's Caldwell catalog became far more popular than any other list except that of Messier's himself. As the reader may already know, the term "Caldwell" was used because in past generations Patrick's family name was Caldwell-Moore, not just Moore, and he could hardly use "M" designations, as these would be confused with the Messier objects he deliberately excluded from his list.

The original Caldwell list, published in that 1995 edition of *Sky & Telescope,* had a number of errors, but all these have now been sorted out and the list has only increased in popularity in the last 15 years. There have been a few critics of the Caldwell catalog in this period (a tiny minority), but since the renowned visual observer Stephen James O'Meara published his massive work on the Caldwell objects most have been silenced. The list was intended purely as a bit of fun, but it caught on like wildfire due to sheer popularity. Even major telescope manufacturers now make telescope hand controllers that include C numbers alongside M numbers.

Chapter 2

The Caldwell Objects

Some Background

For the newcomer to astronomy or deep sky observing, the different types of objects visible through a telescope may be rather confusing at first, so a brief explanation of how these objects fit into the great scheme of things may be useful. Stars are, rather obviously, the brightest objects in the night sky. Their nuclear reactions, created by fusion at their cores, illuminate the darkness. In every case, this fusion process is a controlled hydrogen bomb, with the control being provided by the immense gravity and the nuclear reactions arriving at an amicable agreement. The result of nuclear fusion illuminates the night sky. However, space is very, very big, and so even the stars only appear as pinpoints in the darkness.

Essentially, stars are pretty much all we have to work with to provide any kind of visual illumination in the night sky. When we look at how many different categories of object are visible in the dark universe (and thus the Caldwell catalog), they all link back to stars in some way. Essentially this catalog, much like the Messier catalog, boils down to star clusters, galaxies, nebulae, and supernova remnants. All of these objects are made up of, or related to, stars in some way. Perhaps you may be able to point at dark nebulae, but we would argue here that a black cloud of dust is just an obstruction!

Within the above classifications there are subdivisions. Star clusters can be split into the nearby open clusters well within our Milky Way galaxy, or more distant globular clusters in our galactic halo. They can contain anywhere from hundreds of stars to millions of stars (for the biggest globular clusters). Galaxies have many subdivisions, too. One can generalize and split them into spirals, ellipticals, and irregular galaxies, but even within the spiral category, there are different subcategories, and they will look very different, depending on whether we see them face-on or edge-on or somewhere in between. Galaxies typically contain anywhere from hundreds of millions to hundreds of billions of stars.

The nebulae in this catalog can be categorized into bright, dark, and planetary examples. Bright nebulae are either reflecting the light from nearby stars or being excited by the radiation from nearby stars. Reflection nebulae appear blue because the interstellar dust reflects blue light more efficiently. Dark nebulae are caused by dust blocking our view of something beyond, either a bright nebula or a starfield.

When we come to planetary nebulae, well, they are rather misleadingly named. They have nothing to do with planets whatsoever! The term has arisen simply because most examples are so small and (relatively speaking) bright that they are

M. Mobberley, *The Caldwell Objects and How to Observe Them*,
DOI 10.1007/978-1-4419-0326-6_2, © Springer Science+Business Media, LLC 2009

reminiscent of planets, if, rather ghostly ones. In fact, planetary nebulae are formed from the expelled outer layers of dying stars. Ultraviolet radiation excites the expelled material to glow, making it visible. However, in cosmological terms they are very shortlived, only tens of thousands of years old, a mere eye-blink in the life of a galaxy.

Supernova remnants are the final category of object, and there is only one example in the Caldwell catalog, even if it is split across two catalog numbers, 33 and 34, the Veil, a network or filamentary nebulae near the bright star 52 Cygni. Of course, supernovae were once stars, too, either supermassive stars that collapsed into themselves or white dwarfs that were tipped over the brink by hydrogen from a secondary companion. Supernova remnants glow because they mark the positions where debris from the explosion interacts with gas in the interstellar medium. Much of the emission from the Veil is from doubly ionized oxygen (OIII). It is a sobering thought that the heavy elements making up the body of this author, and you, the reader, were all hurled into the universe via supernova explosions. Without supernovae there would be a lot of hydrogen in the universe but little heavy stuff from which you can make planets, humans, computers, and books.

Some Very Long Distances

In terms of distance from us, and therefore how long the light reaching us has traveled, the Caldwell objects span a huge range of light-years. The closest object in Patrick's catalog is the famous naked-eye Hyades Cluster at 150 light-years distant. We are seeing this cluster as it was during the Victorian era (even if, since Einstein, we have to conclude there is no such thing as absolute simultaneity across the curved time-space universe). A few other Caldwell objects are closer than 600 light-years, specifically C63 (NGC 7293, the Helix Nebula in Aquarius), C68 (NGC 6729, the nebula in Corona Australis), C85 (IC 2391, the Omicron Velorum Cluster in Vela), and C102 (IC 2602, the Theta Carinae Cluster). However, the Hyades Cluster is easily the nearest. As we probe further out from our local region of the Milky Way, we encounter the most distant globular clusters within our galaxy's gravitational influence, and then, after that, we are looking back millions of light-years. As we travel away from our "Local Group" of galaxies we start putting tens of millions of light-years between our spaceship and Earth, but the universe is much bigger than even this. Caldwell 24 (at 230 million light-years) and Caldwell 35 at (300 million light-years) are the largest (200,000 and 250,000 light-years across) and most distant objects in the Caldwell catalog. When the light we see left these galaxies, Earth was in the early Triassic and early Permian/late Carboniferous eras. When you observe deep sky objects you not only peer deep into space, you peer deep into time as well. Of course, the visible universe itself is many billions of light-years across, but beyond a few hundred million light-years even the most energetic galaxies become less than spectacular through an amateur telescope. So, a distance of 300 million light-years marks the limit of the Caldwell catalog – a distance of 2 million times further than the closest Caldwell object, the Hyades.

Explanation of Terms

For each of the objects in the catalog, there is provided a panel of salient data containing our best-guess statistics for each object. Astronomy is not an exact science, at least not when quoting distances and sizes in light-years. No one has travelled to any of these objects, and so, especially for the furthest galaxies, their distances are just an educated guess, and you may well see different figures quoted elsewhere. This uncertainty also applies when quoting the ages of globular clusters. As mentioned in the preface, even in the 1990s astronomers quoted the ages of these ancient objects as being 15, 16, or 17 billion years. It was accepted that this was as old as, or even older than, the accepted age of the universe, but most cosmologists assumed that in future years the age of the universe and the age of the globular clusters would shift to make the universe much older than the oldest globular clusters. In fact, in recent years, a mounting body of evidence indicates the universe is only about 13.7 billion years old, and so the globular clusters must be much younger than this figure.

The first figure quoted for each Caldwell object is the visual **Magnitude**. Of course, these objects are not stars and so, in practice, the fuzzy object can only be compared with a defocused star of a known magnitude to make any meaningful estimate – not the most scientific approach, even if most visual comet observers swear by the method! Observers at dark sites will see a more obvious object, though, as light pollution will make the dimmest parts of the object totally disappear. The **Size** of each object is displayed in degrees, arcminutes, or arcseconds, but, once again, how dark your observing site is, and what aperture you use, will tend to confuse the issue for visual observers. The **Mag/Sec** figure gives the surface brightness of the object in question, namely the magnitude per square arcsecond.

For faint fuzzy objects, such as nebulae or galaxies, the surface brightness can be far more relevant than the total magnitude. The surface brightness figure itself is just as vague but will give you a good idea of how ghostly the object may be. Some large nebulous objects, such as the North American Nebula in Cygnus, seem, at first glance, to have a naked eye magnitude, but their size is significant such that the surface brightness may not be much above that of your local light-polluted sky background. Of all the objects in the Caldwell catalog, the planetary nebulae have, by far, the brightest mag/sec values, but they are often so tiny that they can be overlooked at low powers (they will look just like yet another background star). As far as open clusters are concerned their appearance in the eyepiece will be dominated by the brightness of the individual stars, and so a surface brightness figure is not applicable.

The coordinates of the object listed in the **RA** and **Dec** lines are self explanatory. Whether an object is **Circumpolar** or not, as seen from your latitude is easy to calculate, but this extra piece of data is provided regardless. The **Starhop** data are largely provided to give the reader who is at least partly familiar with the night sky an intuitive feel for where to find the object, by providing offsets from a nearby naked-eye star. In general we have chosen easy stars, even if some are a long way from the target. The reasoning behind this is that only beginners will need this data anyway, as the more experienced Caldwell hunter will know how to chase objects down. Offsets from bright stars to objects are given in degrees of arc; for declination this is straightforward, but for right ascension some further explanation is needed. On the celestial equator (zero degrees declination) one hour of R.A. translates into 15°, and 4 min of R.A. is therefore equal to 1°. Put another way, subtract the offset star's R.A. in minutes from the object's position, divide by 4, and that is the celestial east-west R.A. offset in degrees. However, at higher (or lower) declinations the R.A. lines converge (reaching a point at 90°N or 90°S). So away from the celestial equator you need to divide by 4 and multiply by the cosine of the declination. This is the strategy we have adopted for the star-hop figures, but if you have a right ascension setting circle on your telescope you may well prefer to use the offset in R.A. minutes rather than the "degrees of arc" figures given here.

The **Best Months** data are pretty much self explanatory; essentially the months when the object is highest above the horizon in the late evening hours is listed. Of course, circumpolar objects will always be above the horizon, and as we head further south the "best months" data become increasingly critical for the northern hemisphere observer. (The reverse argument obviously applies when moving north for the southern hemisphere observer). As already mentioned the **Distance** and also the **Physical Size** of the object are astronomer's best guesses and not precise values. Obviously the latter value is totally dependent on the former, as **Distance** × tangent (apparent size) = **Physical Size**. Similarly the **Age** figure is another rough guess made by various astronomers, and the ages quoted can be different, depending on which research report you last read. I should stress that for each object the age quoted ignores any light-travel time to the object in question. If it is stated that a planetary nebula is 10,000-years old but is also 3,500 light-years distant, then we are obviously seeing the nebula 3,500 years late. Therefore, assuming nothing catastrophic has happened to the nebula in the last 3,500 years, it still exists and is actually 13,500-years old. But without constructing a wormhole or building a spacecraft capable of travel in hyperspace we can only comment on its current appearance and so state its age as 10,000 years relative to our current view of it. Planetary nebulae are relatively short-lived phenomena, but by deducing the speed of the ring of debris belched out from the dying star an approximate age can be estimated. As we saw earlier, globular clusters are the oldest objects in the Caldwell catalog, but deducing their ages purely from stellar cosmology can produce figures that challenge the very age of the universe.

The **Best Visual Aperture** data is our "finger in the air" stab at summarizing such a vague preference. It could be argued that, for many deep sky objects, the bigger the telescope aperture the better. So for many objects the minimum aperture to see the object well is stated. However, in the case of large objects of half a degree or more in width, a big aperture, and/or high power can indeed be a disadvantage. But an additional factor to be considered alongside the aperture is which filters to use. Certainly, some narrow band filters, such as the UHC and OIII examples, really do dramatically increase an object's visibility far more than a slight aperture increase. The **CCD/DSLR Advice** simply summarizes the best imaging strategy in one line and the remaining **Celestial Neighbors** and **Miscellaneous** rows are self explanatory. The latter is simply a bit of miscellaneous entertainment for the reader.

The Charts

Many amateur astronomers still use detailed paper star charts appropriate to their instrument to find the field of a faint variable star or an exceedingly faint deep sky object. However, where relatively bright deep sky objects are concerned, these will be very obvious anyway when you sweep around the field with a Dobsonian. Of course, many amateurs simply use "Go To" or old-fashioned setting circles to get within a degree or two of the field (for more on the subject of setting circles see the Chap. 3) If the object is not immediately obvious, they just sweep around. So, in this book we have included basic charts to tell you roughly where to look and plan which object to hunt down, but a detailed star atlas, such as Uranometria 2000, is highly recommended. All of the objects can be pinned down with old-fashioned setting circles and the R.A. and Dec of the object, relative to a nearby star, anyway. An inexpensive software-based planetarium package such as Project Pluto's Guide 8.0 (see the Resources appendix at the end of this book) can be invaluable for printing your own customized charts. Please note, where there is no constellation map on the pages in question the final **Chart** row in the data summary for each Caldwell object will tell you where to find the relevant finder chart on the other Caldwell object pages. Like the images, all the charts have north at the top. There are, admittedly, a few cases where the imager (or this author) has had to orient the field at an angle and in those cases a north arrow has been added, or the orientation is mentioned in the caption. For example, Robert Gendler's excellent image of the Double Cluster in Perseus (Caldwell 14) was rotated anticlockwise so it would fit the page better.

Caldwell 1

NGC 188
Feeble open cluster in Cepheus
Magnitude: 8.1
Size: ~14'
Mag/Sec: Not relevant
RA: 00 h 47 min 30 s
Dec: +85° 15′
Circumpolar: Above 5°N
Starhop: 7½°N and 2½°E from Alrai (γ Cep)
Best Months: All year from northern US/Europe
Distance: 4,800 light-years
Physical Size: 20 light-years
Age: 5 billion years
Best Visual Aperture: Gets interesting above 10 cm
Best Visual Filter: No filter required
CCD/DSLR Advice: Easy object
Celestial Neighbors: Variable RU Cep lies 45′ east
Miscellaneous: One of the most ancient open clusters

Let us not be under any illusions about Caldwell 1. Frankly, it is not going to blow your mind when you have it in the eyepiece. Arguably the biggest attraction of this open cluster is simply its novelty value. It is the only open cluster anywhere near Polaris, but, frankly, if it was not the first Caldwell object many casual deep sky observers would be happy to give it a miss. When this author observed this cluster in October 2008 Stewart Moore, the deep sky section director of the British Astronomical Association, told me there were no other observations of it in the BAA records! However, NGC 188 is dead easy to find even if you need a 10 cm aperture to start to make it vaguely interesting. Everyone in the northern hemisphere can find Polaris, and, from there, you just move 4° south toward the "W" of Cassiopeia. Alternatively starhop from Alrai as suggested in the table. If you continue into Cassiopeia you will find more open clusters than you can shake a stick at, against the backdrop of the Milky Way.

Cosmologically, NGC 188 is a very old open cluster and well above the galactic plane. Messier obviously did not recognize it as a worthy object, even though M26 in Scutum has a similar magnitude and size. However, NGC 188 contains mainly 13th and 14th magnitude stars only obvious as individual pinpoints in 25 cm and larger telescopes.

From an imager's perspective a half-degree-wide field is recommended, as this will permit the inclusion of a number of distinctive 8th magnitude field stars in the frame. Being a cluster with no nebulosity, filtered work is of little benefit. A stack of 30-s DSLR shots will reveal all that the cluster has to offer.

The variable star RU Cephei lies three-quarters of a degree east of NGC 188. RU Cep is a semiregular giant star with a range of 8.2–9.8 and a period of 109 days.

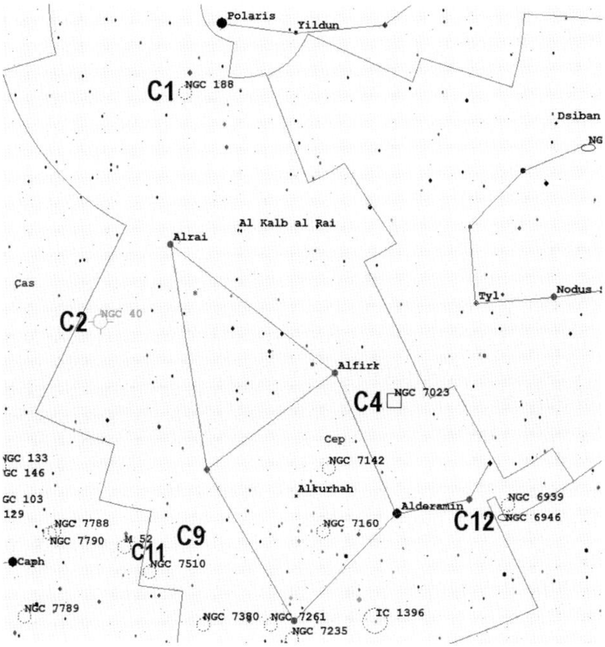

NGC 188 (C1) can be found only 5° south of Polaris in northern Cepheus. Five other Caldwell objects are labeled, four in Cepheus and one just over the border in Cassiopeia.

NGC 188 imaged by the author on 2008 October 26 with a 0.35 m Celestron 14 at *f*/7.7 on a Paramount ME mounting. This was a 120 s exposure with an SBIG ST9XE. The field is 13 arcmin wide.

Caldwell 2

NGC 40
Bow-Tie Nebula in Cepheus
Magnitude: 12.3
Size: 40″
Mag/Sec: 20
RA: 00 h 13 min 00 s
Dec: +72° 31′
Circumpolar: Above 18°N
Starhop: 5°S and 2°E from Alrai (γ Cep)
Best Months: August to November
Distance: 3,500 light-years
Physical Size: 8 light-months
Age: ~4,000 years
Best Visual Aperture: 30 cm and larger with high power
Best Visual Filter: UHC or deep sky filter
CCD/DSLR Advice: Tiny, but high Dec. means easy long exposures
Celestial Neighbors: Faint open cluster NGC 7762, 6°SSW
Miscellaneous: NGC 40 will fade away in 30,000 years
Chart: See Caldwell 1

Travel roughly 13° south of Caldwell 1, passing close by the 3rd magnitude star Alrai at the half way point, and you will arrive in the vicinity of Caldwell 2, the faint planetary nebula NGC 40. Mention planetary nebulae to any northern hemisphere amateur, and the Ring Nebula (M 57) in Lyra will spring to mind. However, that famous celestial showpiece is a beacon compared with the 12th magnitude NGC 40. Conversely, though, NGC 40's central star is much brighter, at around magnitude 11, and tends to swamp the nebula completely below 30×. A decent aperture and high magnifications (at least 200×) will be needed to get the most out of this faint planetary which, at most, has a width of 40 arcsec with bright arcs only at the eastern and western edges, hence the "Bow-Tie" nickname. Actually, this nickname is a bit confusing, as NGC 2440 in Puppis is also called "The Bow-Tie." To some, NGC 40 more resembles the steering wheel from a Formula 1 car. An Ultra High Contrast (UHC) filter will enhance the nebulosity. NGC 40 is a challenging object to image simply because of its small size. Fine sampling (finer than 2 arcsec per pixel) in good seeing will bring the best results. However, poor tracking and less-than-perfect focusing will show up all too easily at these image scales. Only an adaptive optics-based imaging system will show the finest nebulosity: a subtle red/pink lattice of detail between the ends of the bow tie shape, surrounding the central star. Considerable image processing skills are needed to avoid saturating this detail into an overexposed blob. Mind you, on the subject of tracking, at 72° dec. the tracking accuracy will be 1/cosine (72) times less critical, or 3.2×, so the imaging prospects are not all bad.

NGC 40 imaged in fine detail by Tom Carrico from the ARGO Observatory in central Oregon during August 2007. An RCOS 10-in. (25 cm) Ritchey-Chrétien was used with an SBIG ST-10XME CCD camera. Luminance: 40 × 5 min. RGB: 12 × 5 min in each color.

Caldwell 3

NGC 4236
Galaxy in Draco
Magnitude: 9.6
Size: 20′ × 8′
Mag/Sec: 24
RA: 12 h 16 min 40 s
Dec: +69° 28′
Circumpolar: Above 21°N
Starhop: 8°N and 7½°E from Dubhe (α UMa)
Best Months: March/April
Distance: 7 million light-years
Physical Size: 40,000 light-years
Age: Billions of years
Best Visual Aperture: 25 cm and larger
Best Visual Filter: Deep sky filter in severe light pollution
CCD/DSLR Advice: High Dec. means easy long exposures
Celestial Neighbors: Galaxy NGC 4125, 4°S
Miscellaneous: Member of the M 81 group

Roughly 4° east of the magnitude 3.8 star Gianfar (λ Draconis) you may stumble across the almost edge-on barred spiral galaxy NGC 4236, allocated the number 3 in Patrick Moore's Caldwell Catalog. It is near enough to the pole "pointer" Dubhe in Ursa Major to use that star as a starting point with a non "Go To" instrument. However, despite this galaxy's size and its magnitude of 9.6, it is a very tricky object, even with a 25 cm Newtonian. Only when you move up to a 35-cm aperture does it become an obvious faint splinter. The reason, of course, is the surface brightness. Spread over more than half a million square arcseconds the galaxy averages out to around magnitude 24/sq. arcsec, only a few percent extra on top of the sky background even from a reasonable rural site (magnitude20/sq. arcsec). Despite being a relatively nearby galaxy no supernovae have yet been discovered in NGC 4236, but its near edge-on angle to us does us no favors in that respect. You have to look twice at images of this galaxy to realize it is a barred spiral due to the oblique viewing angle and has rather faint spiral arms.

Caldwell 3 is far more rewarding for the CCD imager, especially when equipped with a cooled quantum efficient monochrome detector enabling a clean, deep luminance image to be obtained. Once again the high declination will lessen tracking errors with a 2.8× improvement compared with an object at 0° dec. However, its 20 arcmin length might warrant a sizable detector when using long focal lengths. Four bright variable stars lie in the immediate vicinity of the galaxy, namely VV and SW Draconis and the suspected variables NSV 5534 and 5518.

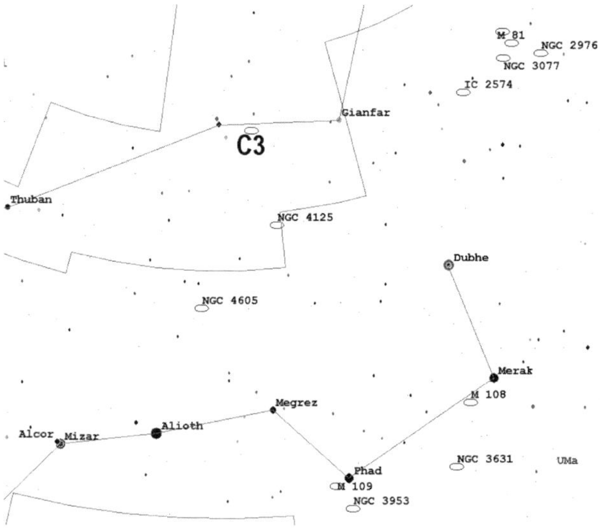

NGC 4236 lives in Draco, roughly midway between the "bowl" of Ursa Major's Plough and the north celestial pole.

NGC 4236 imaged by Tom Harrison from Asheville, North Carolina, USA on April 28/29 2006 using a 12.5-in. (318 mm) *f*/9 Ritchey-Chrétien by RC Optical Systems (RCOS) and an SBIG STL6303E CCD camera. LRGB image 240:60:60:60 with RGB frames binned 2 × 2. Autoguided using a Paramount ME.

Caldwell 4

NGC 7023
Iris nebula in Cepheus
Magnitude: 7
Size: ~20'
Mag/Sec: 22
RA: 21 h 01 min 30 s
Dec: +68° 10′
Circumpolar: Above 22°N
Starhop: 5½°N and 2°W from Alderamin (α Cep)
Best Months: N. hemisphere autumn
Distance: ~1,400 light-years
Physical Size: 8 light-years
Age: ~5,000–6,000 years since star birth
Best Visual Aperture: 10 cm and larger
Best Visual Filter: Deep sky filter in severe light pollution
CCD/DSLR Advice: High Dec. means easy long exposures
Celestial Neighbors: NGC 6951, 2°SW; Gyulbudaghian's var. neb ~1°W
Miscellaneous: Lies 1°WSW of Mira star T Cephei
Chart: See Caldwell 1

Contrary to a number of sources NGC 7023 is a reflection nebula and not a star cluster. At its heart is the magnitude7.4 star V380 Cep, also known as GSC 44603015, HD 200775, SAO 19158, or HIP 103763. The star lies at a distance of 1,400 light-years, give or take about 400 light-years (or ±30%), and has a luminosity roughly 180 times that of our Sun, if it lies at the best-guess distance. V380 Cep is a brand new star that probably ignited as recently as 5,000 or 6,000 years ago. The nebula consists of dust particles in the star's local region that reflect the light of the new star. Of course, if we are seeing the system as it was 1,400 years ago, the light travel time plays a significant part in defining what we mean by the birthdate of the star. Nevertheless, if our ancestors at the dawn of the Bronze Age had possessed telescopes, they might have seen this star turning on! Visually this nebula is quite a treat, especially from far northern latitudes such as the UK, where it can transit at more than 70° altitude on autumn nights. A wealth of intricate, almost fluffy, detail is visible in large instruments even to the novice.

Of course the true colors can only be revealed in color CCD images that show the predominantly blue color characteristic of dust grains reflecting starlight. Closer to the central star two pinky nebular wisps can be seen where dust grains are converting ultraviolet radiation to redder wavelengths by photoluminescence. To avoid overexposing the brighter central regions of NGC 7023, averaging a series of images and intelligent use of a gammalog or DDP image processing routine is recommended. While in Cepheus a short hop 2°SW to the galaxy NGC 6951 is worthwhile, or a hop 1°W to one of only five variable nebulae in the sky (including C 46): Gyulbudaghian's Nebula (20 h 46 min +67° 58′).

NGC 7023 imaged by Gordon Rogers of Long Crendon, Aylesbury, UK with a 0.4 m RCOS Ritchey Chrétien on a Paramount ME and an SBIG ST10XE, CFW8 filter wheel and AO7 adaptive optics unit. http://www.gordonrogers.co.uk

Caldwell 5

IC 342
Galaxy in Camelopardalis
Magnitude: 8.4
Size: 16′
Mag/Sec: 23
RA: 03 h 46 min 50 s
Dec: +68° 06′
Circumpolar: Above 22°N
Starhop: 4½°N and 12°W from Segin (ε Cas)
Best Months: N. hemisphere winter
Distance: ~10 million light-years
Physical Size: ~50,000 light-years
Age: Billions of years
Best Visual Aperture: 10 cm and larger
Best Visual Filter: Deep sky filter in severe light pollution
CCD/DSLR Advice: High Dec. means easy long exposures
Celestial Neighbors: Open cluster NGC 1502, 6°S
Miscellaneous: Discovered by W.F. Denning in the 1890s

The second galaxy in the Caldwell catalog is yet another large, bright, and high declination galaxy missed by Messier. In this case the earliest known reference to it appears to be by the British observer W. F. Denning, in the early 1890s. One can only assume that Messier's obsession with comet hunting, especially at small solar elongations, would make a northern hemisphere winter galaxy at dec. +68 too far off the beaten track (the winter Sun can sink to Dec –23½). At the time of this writing no supernovae have ever been discovered in IC 342. Sod's law therefore guarantees that one will occur days after final proofs of this book have been checked! A face-on nearby spiral like this one is a prime target for supernova hunters. Visually, Caldwell 5 looks less impressive than you might expect for an 8th magnitude galaxy. The source of this problem could be the fact the spiral arms are so ghostly when compared with the galaxy core. However, on a crystal clear winter's night with the object almost overhead, it is a fine sight even in a 16-cm aperture Takahashi Epsilon astrograph at low powers.

For the galaxy imager with a 16 arcmin field IC 342 is a great target, but you need long, deep exposures under really dark skies to show those arms well. In UK, January skies this galaxy transits at 8 p.m., making it ideally placed for those clearest of all winter skies when the CCD camera temperature is close to freezing even before you switch the cooler on. This is one galaxy where the LRGB technique is essential for color images: only an unfiltered luminance image will pull the spiral arms out of the murk.

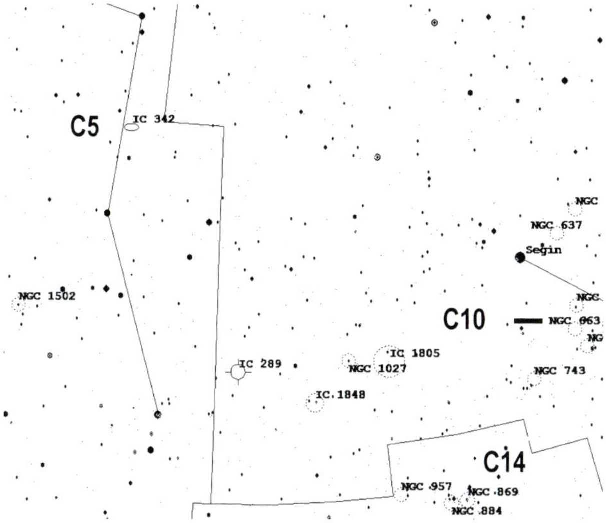

The chart shows the Camelopardalis/Cassiopeia/Perseus border region and the locations of Caldwell objects 5, 10, and 14.

IC 342 imaged by Gordon Rogers of Long Crendon, Aylesbury, UK with a 0.4 m Schmidt-Cassegrain at *f*/6.3 plus SBIG ST8E CCD and AO7 adaptive optics unit. http://www.gordonrogers.co.uk

Caldwell 6

NGC 6543
Cat's Eye Nebula in Draco
Magnitude: 8.2
Size: 23″ × 18″ (inner); 390″ (outer)
Mag/Sec: 15
RA: 17 h 58 min 33 s
Dec: +66° 38′
Circumpolar: Above 24°N
Starhop: 15° due N from Etamin (γ Dra)
Best Months: N. hemisphere summer
Distance: ~3,000 light-years
Physical Size: ~15 light-weeks (inner); 5 light-years (outer)
Age: ~1,000 years
Best Visual Aperture: 20 cm and larger with high power
Best Visual Filter: OIII and UHC filters work well
CCD/DSLR Aadvice: Tiny, but high Dec. means easy long exposures
Celestial Neighbors: Galaxy NGC 6503, 3½°N
Miscellaneous: Nebula lies close to N. ecliptic pole

The Cat's Eye Nebula shares with others the highest surface brightness among Caldwell objects (shared with Caldwell's 15, 22, 55, 59, and 90), and the Hubble Space Telescope (HST) image of this object is as well known as it is stunning. With its inner dimensions of only 23 by 18 arcsec the Cat's Eye appears smaller than Mars at its largest, and so high magnifications are required. By high we mean crazy powers; with a 20 cm aperture try 500 or 600×. Planets cannot bear such powers, but with dark adapted vision your visual resolution is low and high contrast is the issue, not sub arcsecond detail. Do not expect to see HST details, though. You will get a hint of a greenish cast to the nebula, but all most observers will see is an eye shape. The central 11th magnitude star is the pupil; then there is a bright ring and a fuzzy outer halo.

Imaging Caldwell 6 is also complicated by its small size. However, there is a much fainter outer halo some 390″ in diameter, which is easy to image if you have dark skies, even if you will not see the intricate inner region HST details. However, to record a poor man's version of the Hubble images is not impossible. With long focal lengths (such as a modern SCT at f/10), a fine image scale (1 arcsec per pixel and finer), and an adaptive optics unit such as the SBIG AO-7, or AO-8 the major details in the iris of the cat's eye can be captured. But it takes experience and patience to achieve. Another approach, which works well with high surface brightness objects such as planetary nebulae, is to take dozens of images of just a few seconds duration and stack and average them in a software package such as *Registax*. With fainter deep sky objects exposures of such short durations would be very noisy, but planetary nebulae can saturate the detector with exposures of minutes at fast f-ratios, so shorter duration images work well.

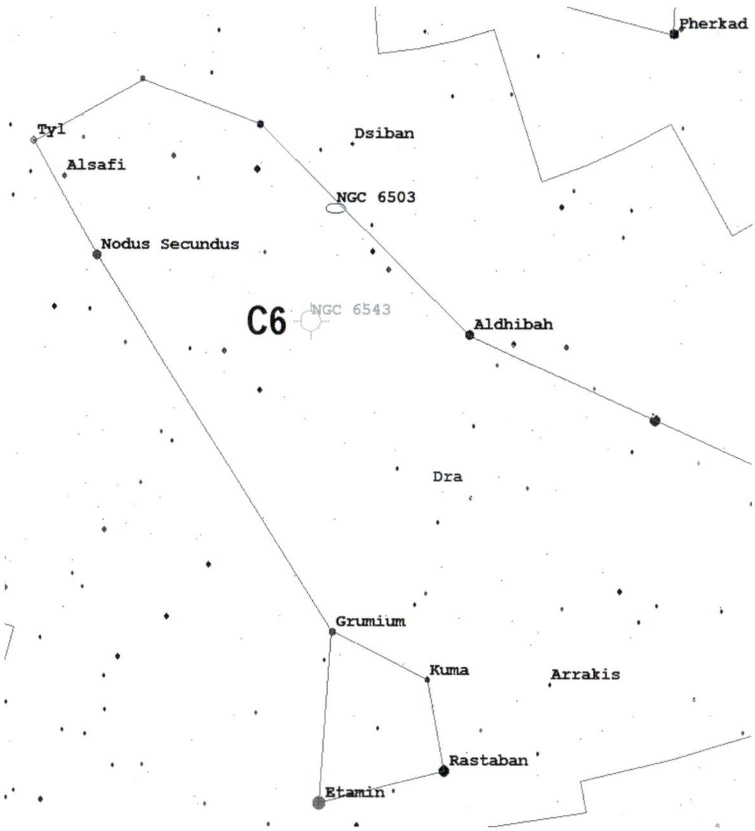

NGC 6543 (C6) lives in central Draco.

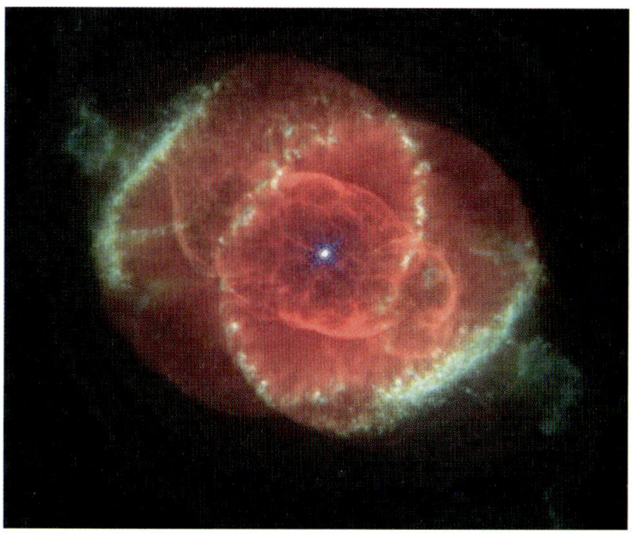

The famous Hubble Space Telescope image of the Cat's Eye Nebula.

Caldwell 7

NGC 2403
Galaxy in Camelopardalis
Magnitude: 8.4
Size: 23′ × 12′
Mag/Sec: 23
RA: 07 h 36 min 50 s
Dec: +65° 36′
Circumpolar: Above 25°N
Starhop: 5°N and 6°W from o UMa
Best Months: N. hemisphere winter and spring months
Distance: ~12 million light-years
Physical Size: ~80,000 light-years
Age: Billions of years
Best Visual Aperture: 10 cm and larger
Best Visual Filter: Deep sky filter in severe light pollution
CCD/DSLR Advice: Easy telescope + CCD target
Celestial Neighbors: M81 and M82 are 13°ENE
Miscellaneous: Produced a superb supernova in 2004

Few galaxies in the night sky are larger in angular size than NGC 2403. Obviously the Andromeda galaxy M31 takes the top spot if the Magellanic clouds are excluded. Then one can think of M33 and M101 and, closer in size, the southern hemisphere's NGC 247 (Caldwell 62), NGC 253 (Caldwell 65), NGC 300 (Caldwell 70), and NGC 55 (Caldwell 72). Perhaps M81 is the best match, though, as it, too, occupies a northern declination in the mid to high sixties. NGC 2403 could easily be mistaken for a bright comet and so, once again, one must think Messier neglected parts of the far northern sky when he was sweeping. Caldwell 7 is a high priority target for supernova hunters. So far it has produced two (2004dj and 2002kg) and a "supernova imposter" (1954J). The imposter, originally designated as a genuine supernova in 1954, is now thought to have been the pre-SN outburst of a star such as the famous eta Carinae. Supernova 2004dj, discovered by the Japanese hunter Itagaki became the 9th brightest in recorded history and one of the best observed. It peaked at around magnitude 11.2 when discovered.

Visually, NGC 2403 is an obvious fuzzy ellipse even in a 160-mm aperture Takahashi Epsilon. Two 10th magnitude stars 4 arcmin apart look tantalizingly like ultra-bright supernovae at first glance but are just Milky Way field stars. For the imager the galaxy is a real treat, showing much detail in the spiral arms even in short exposures. The galaxy's size seems to almost quadruple in the longest exposures, where the full 23′ × 12′ dimensions are captured. In color images, a steely blue spiral is resolved with subtle flecks of pink hydrogen visible in star-forming H II regions.

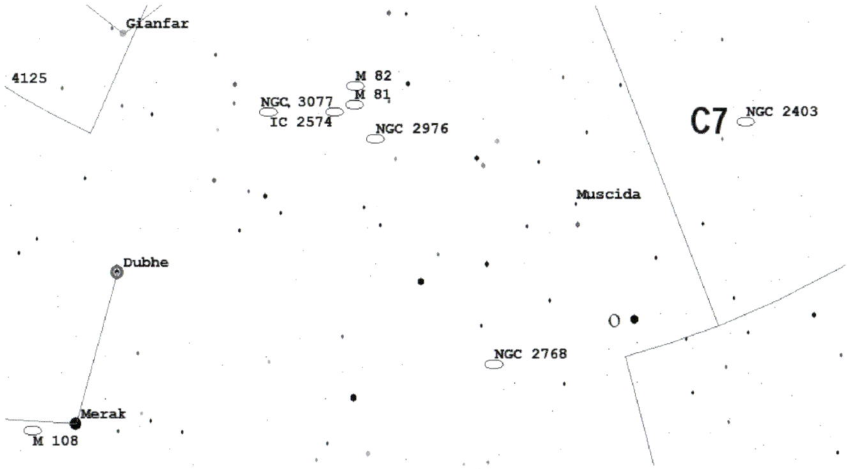

NGC 2403 (C7) lies just west of the Ursa Major/Camelopardalis border, within a celestial stone's throw of Messier 81 and 82.

NGC 2403 imaged by Robert Gendler using two Ritchey-Chrétien telescopes: a 12.5-in. (318 mm) for the RGB color data and a 20-in. (508 mm) for the luminance data, plus RCOS Field Flatteners. The 20-in. mounting was a Software Bisque Paramount ME and the cameras were SBIG STL11000XM units. A total of 15 h of exposures were combined to create this image. The 20-in. RC is located at the New Mexico Skies "Nighthawk" observatory. © Robert Gendler.

Caldwell 8

NGC 559
Open cluster in Cassiopeia
Magnitude: 9.5
Size: 7.0′
Mag/Sec: Not relevant
RA: 01 h 29 min 30 s
Dec: +63° 18′
Circumpolar: Above 27°N
Starhop: 3°N and ½°E from Rucbah (δ Cas)
Best Months: September to January in N. hemisphere
Distance: 3,700 light-years
Physical Size: 8 light-years
Age: ~2 billion years
Best Visual Aperture: 10 cm and larger
Best Visual Filter: No filter required
CCD/DSLR Advice: Dead easy target
Celestial Neighbors: Clusters NGC 609, 637, 654, C10, and M103
Miscellaneous: Supernova remnant G127.1 + 0.5 lies *behind* the cluster

NGC 559 is the second open cluster in the Caldwell catalog. It lives in Cassiopeia, which, frankly, is already stuffed to the bursting point with open star clusters – more than you can shake the proverbial stick at. The best-known examples in this constellation are M52 and M103 at seventh and sixth magnitude, with the rest having NGC or IC designations. As we shall shortly see Patrick credited two further open clusters in Cassiopeia with Caldwell designations (numbers 10 and 13), and the famous double cluster in Perseus (Caldwell 14) is a stone's throw away, too. NGC 559 is sometimes called the Ghost's Goblet. As with Caldwell 1, Caldwell 8 is less than mind blowing, but it is very easy to find. Anyone in the far northern hemisphere can locate the W of Cassiopiea, and NGC 559 forms the northwestern right angle of a triangle comprising itself and the two westernmost W stars, Ruchbah and Segin. Do not be fooled by the brighter open cluster NGC 637 or the fainter NGC 609 in the same vicinity; NGC 559 is much further west of Segin. Visually the densest part of the cluster is less than 5 arcmin across, with most of the brightest stars forming a crescent shape along the eastern edge. The brightest four stars are between magnitude 11 and magnitude 12, with the next brightest dozen only being between magnitude 13 and magnitude 14. In small apertures the stars that are too faint to individually detect look a bit like a ghostly vapor emerging from a cup, to some observers at least. Through a large aperture the fainter stars can be held steady with averted vision.

For the CCD imager the cluster will not excite either, but even the shortest exposure will record what you can see.

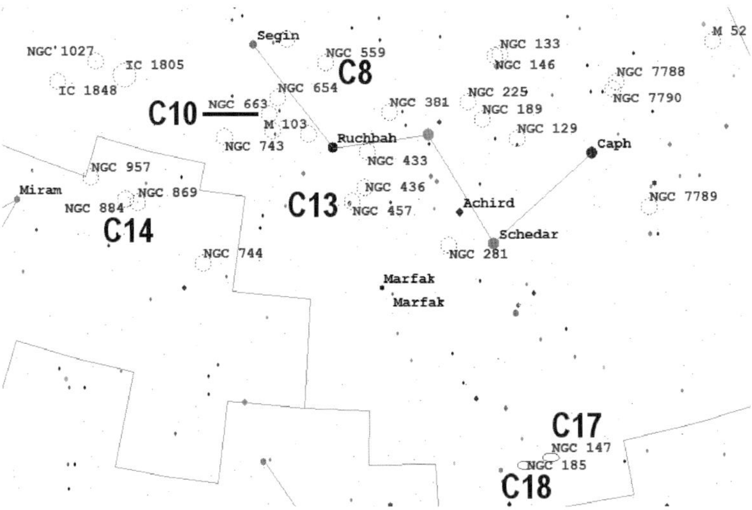

NGC 559 and five other Caldwell objects in Cassiopeia and Perseus.

NGC 559 from the Digitized Sky Survey. The field is 10 arcmin wide.

Caldwell 9

Sharpless 2-155
Cave Nebula in Cepheus
Magnitude: ~8
Size: ~45′ × 50′
Mag/sec: 24–25
RA: 22 h 56 min 50 s
Dec: +62° 35′
Circumpolar: Above 28°N
Starhop: 3½°N and 9°W from Caph (ß Cas)
Best Months: August to December in N. hemisphere.
Distance: ~3,000 light-years
Physical Size: ~40 light-years
Age: Precise age unknown
Best Visual Aperture: 40 cm and larger, if possible!
Best Visual Filter: UHC filter works well
CCD/DSLR Advice: LRGB with H-Alpha luminance is recommended
Celestial Neighbors: C 11 (Bubble) and M52, 5°E
Miscellaneous: Arguably the trickiest visual Caldwell object
Chart: See Caldwell 1

The Cave Nebula is a serious test for CCD imagers and visual observers alike. For imagers the challenge is to record the subtle beauty of this feature, whereas for visual observers the challenge is to see anything at all! For UK imagers our high northern latitude is a real bonus for this feature, enabling it to transit within 10° of the zenith on autumn nights and early winter evenings. In the original Caldwell listing published in Patrick's article in the December 1995 *Sky & Telescope* Caldwell 9 was listed as Sh (Sharpless) 2-155 and named as the Cave Nebula. As Stephen J. O'Meara pointed out in his *Deep Sky Companions –The Caldwell Objects* tome a completely different feature in Cepheus was originally called the "Cave Nebula in Cepheus," but in the modern era the Caldwell Cave has become better known. The Caldwell Cave is in fact a bright H II emission nebula that curves around in an arc allegedly resembling the mouth of a cave. The nebula is part of the much larger Cepheus B molecular cloud. Hot young stars in the region illuminate and ionize the hydrogen gas, making it glow. The word "bright" is, of course, a relative term, and few visual observers could describe the emission nebula responsible for the Cave as bright. You may not see it at all. For those with setting circles starhopping 3½° north and 72 R.A. minutes west from Caph (ß Cas) will get you to the field. Alternatively, find the open cluster M52 and hop 3½° west and 1° north. Seeing what is there is another matter, but as big an aperture as possible, dark and crystal clear skies, and a UHC filter will help. For the CCD imager, a very deep luminance image coupled with R, G, and B images binned 2 × 2 or 3 × 3 will record impressive detail, but for the highest contrast an H-alpha filter and long autoguided exposures for the luminance component, will reveal the most subtle details.

The Cave Nebula in Cepheus imaged by UK astro-imager Ian Sharp of Ham, West Sussex. An 80 mm semi-apochromatic Orion ED80 refractor (600 mm focal length, *f*/7.5) mounted on an Astrophysics AP900 GTO was used. The luminance data (11 × 10 min) was shot through a Hydrogen-alpha filter. ATIK 314L CCD camera.

The Cave imaged by the author on 2008 October 24 at a longer focal length, with a Celestron 14 at *f*/7.7 on a Paramount ME mount and with an unfiltered SBIG ST9XE CCD camera. Four 180-s exposures were stacked to produce this image. Note the huge number of stars that emerge in this unfiltered shot (12.5 arcmin wide).

Caldwell 10

NGC 663
Horseshoe cluster in Cassiopeia
Magnitude: ~7
Size: 15′
Mag/Sec: Not relevant
RA: 01 h 46 min 20 s
Dec: +61° 13′
Circumpolar: Above 29°N
Starhop: 1°N and 2½°E from Rucbah (δ Cas)
Best Months: September to January in N. hemisphere
Distance: 7,200 light-years
Physical Size: ~30 light-years
Age: ~20 million years
Best Visual Aperture: 10 cm and larger
Best Visual Filter: No filter required
CCD/DSLR Advice: Easy imaging target
Celestial Neighbors: Clusters M 103, NGC 659, 654
Miscellaneous: Mysteriously, just as good as M103!
Chart: See Caldwell 8

As mentioned earlier, Cassiopeia has a lot of open clusters, and Caldwell 10, or NGC 663, is the next on the list. However, unlike the rather dismal Caldwell 8 (and Caldwell 1 in Cepheus), this one is worth getting out of bed for; hmm, O.K., worth delaying your beauty sleep for may be more accurate. It is actually quite a bright open cluster at roughly magnitude 7, slightly brighter than M103 and similar to M52 in the same constellation. Once again the eastern "W" stars Ruchbah and Segin can guide you there as the cluster lies just east of the mid-point of the imaginary line joining these two stars. The problem is it is all too easy to get led astray by the other open clusters, as the smaller NGC 654 is barely a degree to the north-northwest. NGC 659 is half a degree to the south-southwest, and M103 is 1° west of the latter. But, to some, Caldwell 10 is the best of this bunch. With a 5°-wide field you can view Ruchbah, Segin, and all these clusters in one go using, say, a 60-mm refractor at 12× and a wide field eye-piece. Why Messier would include NGC 581 (M103) in his catalog but omit the brighter NGC 663 (Caldwell 10) a degree away must remain a mystery. At 50× and above Caldwell 10 is a splendid sight although, for many, the cluster does not instantly advertize its "horseshoe" name. Four cluster stars of 8th and 9th magnitude, in two groups of two, catch the eye immediately, and two faint stellar beards trail south from each. The eastern beard seems to be the most obvious. For those with a vivid imagination the two beards link to form a horseshoe. Through large apertures a hundred stars brighter than magnitude 14 can be glimpsed.

NGC663 imaged by Robert Gendler using a 12.5-in. (318 mm) *f*/9 Ritchey-Chrétien (by Optical Guidance Systems) and SBIG STL11000 CCD camera. Astrophysics AP1200 mount. LRGB image with 15 min each in L, R, G and B, i.e. LRGB = 15:15:15:15. © Robert Gendler.

Caldwell 11

NGC 7635
Bubble Nebula in Cassiopeia
Magnitude: ~10 (NGC 7635 total)
Size: 15′ × 8′
Mag/Sec: ~24
RA: 23 h 20 min 40 s
Dec: +61° 12′
Circumpolar: Above 29°N
Starhop: 2°N and 6°W from Caph (ß Cas)
Best Months: August to December in N. hemisphere.
Distance: ~7,000 light-years
Physical Size: ~6 light-years (bubble)
Age: ~300,000 years
Best Visual Aperture: 40 cm and larger if possible!
Best Visual Filter: OIII or UHC
CCD/DSLR Advice: LRGB with deep luminance works well
Celestial Neighbors: Clusters M52, NGC 7510; C9 (Cave)
Miscellaneous: Central star contains ~15 solar masses
Chart: See Caldwell 1

If you know how to locate the open cluster M52 then finding Caldwell 11 is easy; its brightest part lies a mere half degree to the cluster's southwest. Some clarification is required in defining this Caldwell object. The moniker "Bubble Nebula" applies to a specific part of NGC 7635, namely, a 3 arcmin diameter ring containing the magnitude 8.7 star GSC 4279 1582. Dramatic images by the Hubble Space Telescope have promoted the Bubble, and its name is often casually exchanged with NGC 7635, even if the nebula is much larger, virtually touching M52 in wide images. The two are, of course, not associated, and M52 is thought to be some 2,000 light-years closer to Earth. The Bubble Nebula is a spherical hole blown in the NGC 7635 emission nebula by radiation pressure from the magnitude 8.7 star. Visually, the field can be quite confusing without a bit of prior homework. As well as NGC 7635 being much larger than the bubble a bright part of the nebula just north of the bubble itself may be mistaken for the famous object. The magnitude 8.7 star within the bubble does observers no favors either, and you will simply wish it was not there. This is an object for big amateur telescopes. Dobsonians in the 40–50 cm aperture range will succeed whereas smaller instruments will fail, especially when equipped with an Oxygen III or UHC filter. With the raw quantum gathering power of a CCD, deep sky imagers can take stunning images of the Bubble Nebula and its surroundings. Three arcminutes is a handy size even for small detectors – large enough for seeing issues to be of secondary importance but easy to position on the chip. A deep luminance image with a deep sky H-alpha filter will boost the contrast on LRGB images, but unmodified DSLRs with their poorer red sensitivity will struggle.

The Bubble Nebula in Cassiopeia imaged by UK astro-imager Ian Sharp of Ham, West Sussex. An 80 mm semi-apochromatic Orion ED80 refractor (600 mm focal length, *f*/7.5) mounted on an Astrophysics AP900 GTO was used. The camera was an ATIK 314L. The luminance data (2×15 min + 1×10 min) was shot through a Hydrogen-alpha filter.

Caldwell 12

NGC 6946
Galaxy on Cepheus/Cygnus border
Magnitude: 8.8
Size: 11′ × 9′
Mag/Sec: 23
RA: 20 h 34 min 50 s
Dec: +60° 09′
Circumpolar: Above 30°N
Starhop: 2½°S and 4½°W from Alderamin (α Cep)
Best Months: July to October in N. Hemisphere.
Distance: ~15 million light-years
Physical Size: ~50,000 light-years
Age: Billions of years
Best Visual Aperture: 20 cm and larger
Best Visual Filter: Deep sky filter may help
CCD/DSLR Advice: Deep unfiltered images work well
Celestial Neighbors: Cluster NGC 6939, 40′ NW
Miscellaneous: Ultra-prolific supernova producer
Chart: See Caldwell 1

For the northern hemisphere supernova patroller NGC 6946 is *the* galaxy to patrol. What an outstanding performer! Nine supernovae have been discovered in this spiral since 1917, and, with three between 2002 and 2008, many must have been missed in the photographic era. Although any unfiltered CCD image will reveal the textbook spiral shape, visually Caldwell 12 is rather disappointing. The surface brightness is low, not helped by the galaxy's position on the Cepheus/Cygnus border and thus affected by Milky Way dust extinction. A large telescope and crystal clear skies with the galaxy near the zenith (easy in UK autumn months) will pay dividends. But, even in a 40-cm aperture, most observers will just see a series of subtle concentric glows and not the spiral form.

Designation	Date	Discoverer	Disc. magnitude	Offset	Type
2008S	Feb 1	Arbour	17.6	53W 196S	IIn?
2004et	Sep 27	Moretti	12.8	250E 120S	II
2002hh	Oct 31	LOTOSS	16.5	61W 114S	II
1980k	Oct 28	Wild	11.4	280E 166S	II-L
1969P	Dec 11	Rosino	13.9	5W 180S	N/A
1968D	Feb 29	Wild and Dunlap	13.5	45E 20N	II
1948B	Jul 6	Mayall	14.9	222E 60N	II
1939C	Jul 17	Zwicky	13.0	215W 24N	N/A
1917A	Jul 19	Ritchey	14.6	37W 105S	N/A

A list of supernovae discovered in NGC 6946 with their discovery dates, discoverers, magnitudes, offsets from the galaxy nucleus in arcseconds, and supernovae types.

The face-on spiral galaxy NGC 6946 and one of its many supernovae, imaged by the author on 2004 October 18 using a Celestron 14 at *f*/7.7 and an SBIG ST9XE CCD. Supernova 2004et, shining at magnitude 13.0 is indicated. The image is the result of three stacked 180 s exposures. Paramount ME mounting.

Caldwell 13

NGC 457
Phi Cas cluster in Cassiopeia
Magnitude: 6.4
Size: 20′
Mag/Sec: Not relevant
RA: 01 h 19 min 30 s
Dec: +58° 17′
Circumpolar: Above 32°N
Starhop: 2°S and 1°W from Ruchbah (δ Cas).
Best months: September to January in N. Hemisphere.
Distance: 9,000 light-years
Physical Size: ~50 light-years
Age: ~15 million years
Best Visual Aperture: 10 cm and larger
Best Visual Filter: No filter required
CCD/DSLR Advice: Easy target
Celestial Neighbors: Clusters NGC 436 and 433 to N
Miscellaneous: Phi Cas. double represents E.T.'s eyes!
Chart: See Caldwell 8

The fourth open cluster in the Caldwell catalog is yet another object not far from Ruchbah in the "W" of Cassiopeia. From that star a hop of 2° south and 1° west will get you there. However, the magnitude 5 star Phi Cas is situated right on the cluster's southeastern edge and so can be seen even through the smallest telescope moving from Ruchbah to Phi. Although the Horseshoe cluster (Caldwell 10) is a nice one, the Phi Cas cluster is arguably the best in Cassiopeia. It is bright and fairly large, and the additional bonus of Phi Cas does not dazzle the observer; indeed, it adds to the spectacle. Perhaps more important is the fact that the cluster includes stars of a range of visual brightnesses; as well as Phi, its magnitude 7 companion (NSV 466), and numerous stars of ninth and tenth magnitude fill the inner 10 arcmin. The central region has an X-shaped core, which the eye tends to interpret in various forms. In the United States, many observers see the extraterrestrial E.T. in there, Phi and NSV 466 marking his eyes. However, it appears likely that Phi and its companion are significantly nearer to Earth and not physically part of the cluster. From an imager's viewpoint any imaging device attached to a modest telescope can acquire a good image of Caldwell 13, assuming the resulting system has a 20 arcmin field. Newtonian users have an extra bonus, though. The vanes supporting a Newtonian secondary will produce a pretty star-shaped diffraction pattern around Phi that will enhance any image without resorting to artificial star shape software trickery.

The Phi Cas cluster imaged by Robert Gendler, using an RCOS 12.5-in. Carbon Truss Ritchey-Chrétien with RCOS Field Flattener/Corrector. Software Bisque Paramount ME, SBIG STL-11000XM. LRGB. 6.5 hours of imaging data acquired at the New Mexico Skies Nighthawk Observatory. © Robert Gendler.

Caldwell 14

NGC 869 and 884
Double Cluster in Perseus
Magnitude: 5.3 and 6.1, respectively
Size: Both approx 18′ diameter
Mag/Sec: Not relevant
RA: 02 h 19 min (869) and 02 h 22 min (884)
Dec: +57° 08′ (both)
Circumpolar: Above 33°N
Starhop: 3°S and 7°E from Rucbah (δ Cas)
Best Months: September to January in N. hemisphere
Distance: 7,300 light-years
Physical Size: Each ~40 light-years
Age: ~5 million years
Best Visual Aperture: Binoculars and beyond!
Best Visual Filter: No filter required
CCD/DSLR Advice: Easy target with a one degree field
Celestial Neighbors: Nothing as impressive
Miscellaneous: NGC 869 = h Per; NGC 884 = chi Per
Chart: See Caldwell 8

Surely every amateur astronomer who has been studying the night sky for a year or more will need no introduction to the double cluster in Perseus. What an awesome spectacle in any telescope, yet, mysteriously absent from the Messier objects despite the Pleiades cluster being allotted the number 45. It would seem Messier did not omit objects just because they were obvious spectacles, so why did he leave out NGC 869 and 884? Frankly, a starhop is not needed to find the Double Cluster; just look midway between the center of Cassiopeia's "W" and Mirphak in Perseus and, even with the naked eye, you will spot a misty patch in the Milky Way. When you look through a telescope and consider that you are looking at two clusters each containing several hundred stars within a 40 light year diameter you conclude that life for an alien deep sky observer in either cluster must be pretty rough. Even on their own each open cluster would merit a high ranking in any deep sky observers list of favorites, but NGC 869 and 884 sit side by side; any closer and they would overlap. The magnitudes of each cluster vary with whatever catalog is used to source the data. The *Deep Sky Field Guide to Uranometria 2000* gives values of 5.3 and 6.1 for NGC 869 (western cluster) and 884 (eastern cluster), but values of almost two magnitudes brighter appear on the Internet. A decent image paints a much clearer story. Remove the brightest stars from each cluster center, and they would look very similar. NGC 869 simply has a collection of magnitude 6.5 to magnitude 8.5 stars at its core that NGC 884 does not have. Magnitude 8.0 and fainter stars occupy the core of this fainter cluster. Color images will reveal two deep red stars between the clusters and a further three on the eastern side of NGC 884 that look strikingly different from the predominantly blue stars in each cluster.

The awesome double cluster in Perseus imaged by Robert Gendler. Cumulative Exposure 14 Hours. Data taken with the 12.5-in. (318 mm) Carbon Truss Ritchey-Chrétien and SBIG STL11000XM plus field flattener/corrector. Image data acquired at the Nighthawk Observatory. North is to the left in this image. © Robert Gendler.

Caldwell 15

NGC 6826
Blinking Planetary in Cygnus
Magnitude: 8.8
Size: 25″
Mag/Sec: 15
RA: 19 h 44 min 48 s
Dec: +50° 31′ 30″
Circumpolar: Above 40°N
Starhop: 5¼°N and 9½°W from Deneb
Best Months: May to November in N. Hemisphere
Distance: ~2,000 light-years
Physical Size: ~3 light-months
Age: ~1,000 years
Best Visual Aperture: 15 cm and larger with high powers
Best Visual Filter: OIII
CCD/DSLR Advice: Tiny, so 1″/pixel and finer works best
Celestial Neighbors: Open cluster NGC 6811, 4°S
Miscellaneous: Discovered by William Herschel in Sept. 1793

Like Caldwell 6, the Cat's Eye Nebula in Draco, Caldwell 15 has a high surface brightness but a small size. Thus, it is easily overlooked at low powers but, once spotted and a high magnification employed, it is a rewarding sight. Its nature is obvious above 100×. NGC 6826, also known as the "blinking planetary," lives above the eastern wing of the Swan. Using setting circles you can hop 5¼° north and 57 R.A. minutes west from Deneb or simply find magnitude 2.9 δ Cyg, the eastern star in the Cygnus Cross, and travel almost 5½° N. Stop here. A distinctive pair of magnitude 6 double stars (16 Cygni and its companion) lie half a degree to the west of Caldwell 15, aiding its identification. The blinking planetary has a star of magnitude 10.6 in its center. If you look directly at the central star at, say, 150×, you will only see the star using your retina's central cones. If you look to one side by the right amount (roughly 8–16° off center) you will see the nebula by using the retina's sensitive rods. You are then successfully using averted vision and have discovered why NGC 6826 is called the blinking planetary. In fact, this trick can be carried out with other planetary nebulae, providing the central star is bright enough. A faint central star will be invisible to the high-resolution color-sensitive cones in the center of the retina. Visually, the larger the telescope the better, and do not be afraid to whack in the highest eyepiece/Barlow/Powermate combination in your collection. Powers of 500 or 600× are quite reasonable when observing an object that looks no larger than Mars at its closest, and O III or UHC filters will increase the contrast. As with so many small deep sky objects, the Hubble Space Telescope reigns supreme in obtaining the finest images of the blinking planetary, but in good seeing and with adaptive optics (like SBIG's AO-7/8 units) splendid amateur images can also be secured.

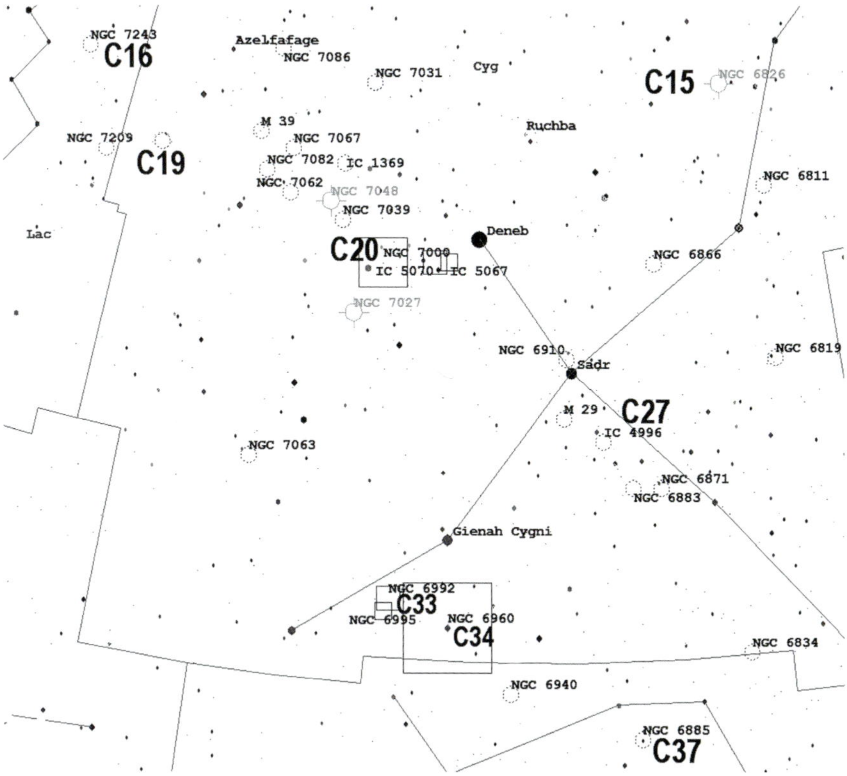

The "Blinking Planetary" NGC 6826 (C15) sits in NW Cygnus and in the *top right* of this finder chart. An additional seven Caldwell objects are shown in Cygnus, Lacerta, and Vulpecula.

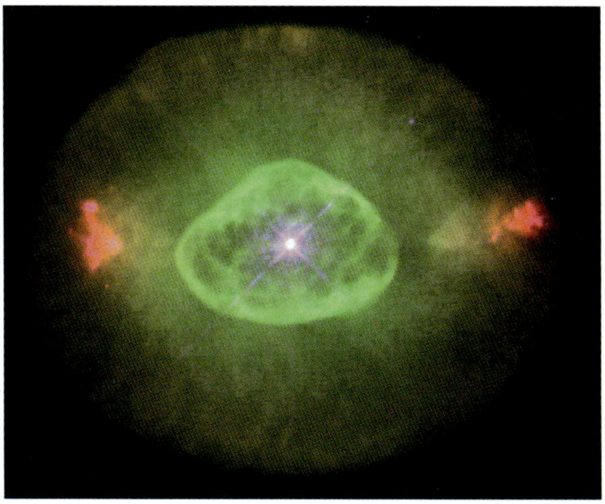

This spectacular close-up of the "Blinking Planetary" nebula was taken with the Hubble Space Telescope.

Caldwell 16

NGC 7243
Open cluster in Lacerta
Magnitude: 6.4
Size: 30′
Mag/Sec: Not relevant
RA: 22 h 15 min 00 s
Dec: +49° 54′
Circumpolar: Above 41°N
Starhop: 4½°N and 16°E from Deneb
Best Months: August to December in N. Hemisphere
Distance: 2,500 light-years
Physical Size: ~20 light-years
Age: 75 million years
Best Visual Aperture: Big binoculars and beyond!
Best Visual Filter: No filter required
CCD/DSLR Advice: Easy imaging target
Celestial Neighbors: Cluster NGC 7209 & C19 (Cocoon) ~3°SSE
Miscellaneous: More an illusion than a cluster?
Chart: See Caldwell 15

Caldwell 16 can, at best, be described as a rather bland Milky Way enhancement in the equally dull constellation of Lacerta, the Lizard. Indeed, one only has to look at a wide field picture of the region to wonder if NGC 7243 exists at all. At this point the reader may well wonder what is going on, questioning the actual existence of a well-established NGC cluster. To get to this potentially nonexistent cluster, you can starhop all the way from Deneb, but it is a long haul. Alternatively, you can find the magnitude 3.8 star α Lacertae and move 2½°W and almost ½°S. You may certainly see something that looks like a cluster in the eyepiece at that point. There is no obvious "edge" to NGC 7243, though, and when you sweep around its environment with a 1° field (and a bit of skepticism) you can come to the conclusion that if you removed half a dozen 9th magnitude stars, including the small triangle at the center (with a double at the southeast vertex), the illusion of a cluster would disappear. Part of the problem here is that the northern half of Lacerta encroaches into the densest part of the Milky Way. In fact, NGC 7243 lies on the same Milky Way isophote that stretches from just north of Deneb. This very high background star density makes it impossible to tell if the cataloged 30 arcmin cluster is being lost within the Milky Way or if it never existed, except in the mind's eye. Perhaps for those who want a real challenge, a more rewarding quarry is the nearby 12th magnitude planetary nebula, IC 5217, 2° to the northeast.

NGC 7243 from the Digitized Sky Survey. The field is 30 arcmin wide.

Caldwell 17

NGC 147
Galaxy in Cassiopeia
Magnitude: 9.5
Size: 18′ × 11′
Mag/Sec: 24
RA: 00 h 33 min 12 s
Dec: +48° 30′
Circumpolar: Above 42°N
Starhop: 8°S and 1°W of Schedar (α Cas)
Best Months: August to December in N. Hemisphere
Distance: 2.5 million light-years
Physical Size: 13,000 light-years
Age: Billions of years
Best Visual Aperture: 25 cm and larger is recommended
Best Visual Filter: Deep sky filter may help
CCD/DSLR Advice: Deep unfiltered images work well
Celestial Neighbors: C18 just 1°E
Miscellaneous: Star formation ceased a billion years ago!
Chart: See Caldwell 8

Caldwell's 17 and 18 are often discussed together, as they are similar-sized dwarf galaxies at the same distance from Earth. Both lie in southern Cassiopeia and are a degree apart in right ascension. Look at a star map and you will see that both galaxies and 11th magnitude NGC 278 (2° further east) sit in a rather unnecessary southern extension to Cassiopeia. If not for this 2° anomaly in the constellation border all three galaxies would be in Andromeda. NGC 51, 3° to the west, has escaped the boundary anomaly. Indeed, the giant Andromeda Galaxy M31 and its two satellites, M32 and M110, lie only 7° to the south, and in many ways these satellites should be considered siblings of Caldwells 17 and 18, as all four are dwarf galaxies within the gravitational influence of the monstrous M31.

NGC 147 is not a pretty galaxy, visually or in images. Despite both galaxies' ninth magnitude status and similar apparent size, NGC 147 is far trickier to spot. This author observed and imaged both these objects on the very clear evening of February 6, 2008, using a 25 cm Newtonian for the visual view and a 35 cm Celestron 14 to take the images. Both galaxies were well past the south meridian but still almost 60° above the western UK horizon. NGC 147 is obviously elongated in a decent aperture, and we view it at an oblique angle. In that respect it is quite similar to M31. In fact there were clearly similarities between NGC 147 through the telescope and M31, 7° lower, through the finder. From an imager's viewpoint, an unfiltered exposure will record the brightest 6 or 7 arcmin length of the galaxy without too much trouble. The full 18 arcmin length is far more elusive, though, and virtually invisible with filtered exposures. A tight cluster of foreground Milky Way stars appears at the core of NGC 147 in any long exposure.

NGC 147 imaged by the author on 2008 February 6 using a 35-cm aperture Celestron 14 at *f*/7.7 and an SBIG ST9XE CCD. This is a stack of two 120 s exposures. The field is 13 arcmin wide. Paramount ME mounting.

Caldwell 18

NGC 185
Galaxy in Cassiopeia
Magnitude: 9.2
Size: 17′ × 14′
Mag/Sec: 23
RA: 00 h 39 min 00 s
Dec: +48° 20′
Circumpolar: Above 42°N
Starhop: 8°S and ¼°W of Schedar (α Cas)
Best Months: August to December in N. Hemisphere
Distance: 2.5 million light-years
Physical Size: 12,000 light-years
Age: Billions of years
Best Visual Aperture: 15 cm and larger is recommended
Best Visual Filter: Deep sky filter may help
CCD/DSLR Advice: Deep unfiltered images work well
Celestial Neighbors: C17 just 1°W
Miscellaneous: First photographed by Keeler (Lick) ~ 1899
Chart: See Caldwell 8

Strangely, Caldwell 17's brother, a degree to its east, appears much more obvious through the eyepiece, despite its similar magnitude and size. Perhaps its more circular aspect is responsible, and one gets the impression that the central region of NGC 185 contains a significant proportion of the galaxy's integrated magnitude of 9.2. Tripod-mounted 80-mm binoculars pick up the glow of NGC 185, but you might be deceived if you think to extend this to the more edge-on NGC 147. If you were sweeping for comets in Cassiopeia, which, frankly, is unlikely, the lack of an obvious spiral structure in NGC 185 and its visual diameter of around 5 arcmin, with a brighter center, might make you think you had bagged a new one. Neither Caldwell 18 nor Caldwell 17 have ever produced a supernova that has been discovered. Of course, over thousands of years they must have produced a few, even given their small size. But when one considers that only one supernova has ever been found in the giant M31, in 1885, at a similar distance, these dwarf galaxies may go millennia without one occurring. In short CCD exposures of NGC 185 a small "eyelid" type of feature becomes obvious to the northwest of the galaxy core. This is, to some viewers, very reminiscent of the "black eye" dust lane seen at the center of M64. In longer exposures, it is all too easy to saturate this feature, so averaging exposures, using a gammalog stretch or the DDP function, is highly recommended. If you get bored with Caldwell 18 (and this is not impossible), a short hop of less than 2° east-southeast will take you to a more challenging 11th magnitude galaxy, NGC 278. In 2004, astronomers Knapen, Whyte, de Blok, and van der Hulst suggested this galaxy had recently interacted with another galaxy now dispersed in the outer disk of NGC 278.

NGC 185 imaged by the author on 2008 February 6 using a 35-cm aperture Celestron 14 at *f*/7.7 and an SBIG ST9XE CCD. This is a stack of two 120 s exposures. The field is 13 arcmin wide. Paramount ME mounting.

Caldwell 19

IC 5146/Collinder 470
Cocoon Nebula and Cluster in Cygnus
Magnitude: ~9 (nebula)
Size: 15′ or 2° (see text)
Mag/Sec: 23 (nebula)
RA: 21 h 53 min 30 s
Dec: +47° 15′
Circumpolar: Above 43°N
Starhop: 2°N and 12½°E of Deneb
Best Months: August to December in N. Hemisphere
Distance: 3,300 light-years
Physical Size: ~15 light-years (excludes B 168)
Age: ~1 million years for Collinder 470
Best Visual Aperture: 15 cm and larger is recommended
Best Visual Filter: UHC and H-beta
CCD/DSLR Advice: LRGB with any field from ¼° to 2°
Celestial Neighbors: C16, 3° NNW & NGC 7209, 2° ESE
Miscellaneous: Disc. By Rev. Thomas Espin in Aug. 1899
Chart: See Caldwell 15

The first time that newcomers to astronomy see an image of the dark snake in Cygnus that meanders its way to the Cocoon (and the associated star cluster) they say "where can I find *that*?" or something similar. The nebula and cluster themselves would be a deep sky attraction, but the river of blackness that worms its way to them makes this feature one of the most photogenic objects in the night sky. Good images of this whole system can be obtained with virtually any medium focal length imaging system. The dark snake is about 2° long, whereas the nebula and associated cluster is roughly 15 arcmin across. In color images, the red nebulosity of IC 5146 is a bit reminiscent of a half-size version of the Trifid Nebula, M20. In higher resolution shots it resembles a fifth-scale Rosette Nebula and its cluster (see Caldwell's 49 and 50). It may be worth clarifying that Caldwell 19's nebula is officially designated IC 5146, the associated cluster is known as Collinder 470, and the dark snake is Barnard 168. The Cocoon lies 2° north and 12½° east of Deneb. Alternatively, you can find the magnitude 4 star ρ Cygni and hop 3½° east and 1¾° north. With any decent aperture the subtle 9th magnitude luminosity of the nebula should appear. However, below 15 cm in aperture it will be disappointing, and the two brightest stars in the associated cluster, 9th magnitude NSV 13937 and NSV 13931 do not make life easy. Perhaps surprisingly the meandering snake is quite easy to see, even in binoculars. It is superbly placed on warm northern latitude August and September evenings when the pleasant temperatures allow you to sweep around east of Cygnus and near the border with Lacerta without fearing death through frostbite!

The Cocoon cluster, nebula and part of the dark rift snaking away from it in the northern Milky Way. Imaged by UK astro-imager Ian Sharp of Ham, West Sussex. An 80 mm semi-apochromatic Orion ED80 refractor (600 mm focal length, *f*/7.5) mounted on an Astrophysics AP900 GTO was used. The camera was an ATIK 314L. LRGB: 10 × 10 min luminance plus 6 × 10 min per color.

Caldwell 20

NGC 7000
North American Nebula in Cygnus
Magnitude: ~5
Size: ~2°
Mag/Sec: 24
RA: 20 h 58 min 50 s
Dec: +44° 20'
Circumpolar: From above 46°N
Starhop: 1°S and 3°E from Deneb
Best Months: July to November in N. Hemisphere
Distance: ~1,800 light-years
Physical Size: ~60 light-years
Age: Precise age unknown
Best Visual Aperture: 10 cm at 20× is optimum
Best Visual Filter: UHC or OIII
CCD/DSLR Advice: Good with fast, 300–500 mm, lenses
Celestial Neighbors: Pelican nebula next door
Miscellaneous: Ionizing star may be Deneb or HD 199579
Chart: See Caldwell 15

Although there are quite a few nebulae that really do resemble their popular names in photographic appearance, NGC 7000 really *does* look like North America. Indeed, if your imagination is vivid enough you can extend the resemblance as far as Costa Rica and Panama. To the west a fainter "dead ringer" can be seen in very wide and deep exposures: IC 5070 really does look like a Pelican, too! The North American/Pelican nebulosity can easily be located with the dark adapted/naked eye roughly 3° east-southeast from the unmistakeable Deneb. However, do not expect it to look like the United States and Canada without optical aid; it is just a subtle brightening of the Milky Way. For the best view, a wide-field instrument with a field of around 3° is advised. For a 5-mm exit pupil, attainable by most dark adapted eyes, and using a quality (60° apparent field) eyepiece a magnification of around 20× with a 100 mm aperture is optimum. With this combination you will be able to see the North American shape emerge. However, the pink glow seen in deep exposures will be invisible, as the retina's low light rods are monochrome detectors. At best it will appear a gray/green color, like so many deep sky objects. Any imager with a system allowing a 2–3° field can record the North American shape of NGC 7000. A 4° field will allow the entire North American/Pelican region to easily be captured. Cooled CCD cameras will do far better than standard DSLRs on these objects due to the insensitivity of unmodified DSLR sensors to the red end of the spectrum. The highest contrast pictures of NGC 7000 are achieved by using deep (multi-hour) CCD exposures using an H-alpha filter for the luminance component.

The North American Nebula (*left*) and Pelican Nebula (*right*) shoehorned into the same field of view by Ian Sharp of Ham, West Sussex. Takahashi Epsilon 130-mm Astrograph (f/3.6) and Hutech-modified Canon 350D on an Astrophysics AP1200 GTO mount. 2007 September 4, 30 × 120 s exposures stacked.

Caldwell 21

NGC 4449
Box Galaxy in Canes Venatici
Magnitude: 9.6
Size: 5.5′ × 4.0′
Mag/Sec: 22
RA: 12 h 28 min 12 s
Dec: +44° 06′
Circumpolar: Above 46°N
Starhop: 5¼°S and 13½°W from Alkaid (η Ursa Major)
Best Months: March to May N. hemisphere
Distance: 10 million light-years
Physical Size: ~16,000 light-years
Age: Billions of years
Best Visual Aperture: 15 cm and larger is recommended
Best Visual Filter: Deep sky filter may help
CCD/DSLR Advice: Deep, unfiltered b/w images work well
Celestial Neighbors: M106 and M94 within 5°
Miscellaneous: Similar object to the Small Magellanic Cloud

Northern hemisphere galaxy hunters will be well aware that the spring months offer an infinity of bright offerings to the deep sky fanatic and supernova hunter alike. Moving north from Virgo into Coma Berenices and onwards through Canes Venatici and into Ursa Major, dozens of bright galaxies become visible even in modest telescopes. Those in the former two constellations rapidly become lost in summer twilight by June (at least from northern latitudes), but many in Canes Venatici and Ursa Major are still accessible in early autumn evenings. Caldwell 21 comes into this latter category, although it is best placed before June; it is actually the northernmost of seven Caldwell galaxies close together near the Canes Venatici/Coma Berenices border, as is obvious from the chart.

To find NGC 4449 without a "Go To" system you can use the tail star Alkaid in Ursa Major or, if you prefer, locate magnitude 2.9 Cor Caroli, the only bright star in Canes Venatici, then hop 5° northwest to 4th magnitude Chara, and finally hop a further 3° north-northwest. Although the whole region is stuffed full of galaxies the nearest brightish galaxy to Caldwell 21 is ¾° further north and more than two magnitudes fainter. If you are already familiar with the Messier galaxies, Caldwell 21 sits midway between M106 and Chara. Although not the prettiest object in deep exposures, this irregular dwarf galaxy is surprisingly easy even in modest apertures, and much structure is visible in really clear spring skies. Of the seven Caldwell galaxies encountered so far on our journey south this one has the highest surface brightness. A quick hop 3° to the north-northwest enables you to compare M106 with Caldwell 21. Although the Messier galaxy is bigger and brighter the surface brightness of the Caldwell galaxy appears brighter (to some at least) in a 25 cm reflector at 200×.

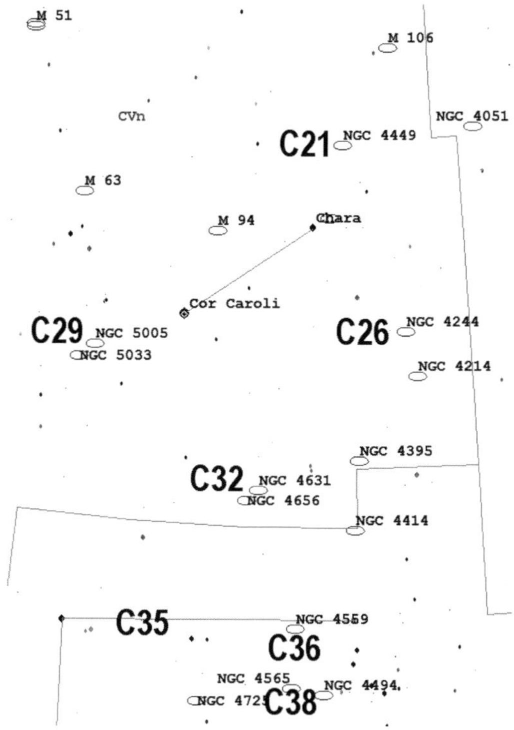

The dwarf galaxy NGC 4449 (C21) in Canes Venatici is shown in the finder chart, along with three other Caldwell galaxies in this constellation and three others over the border in Coma Berenices.

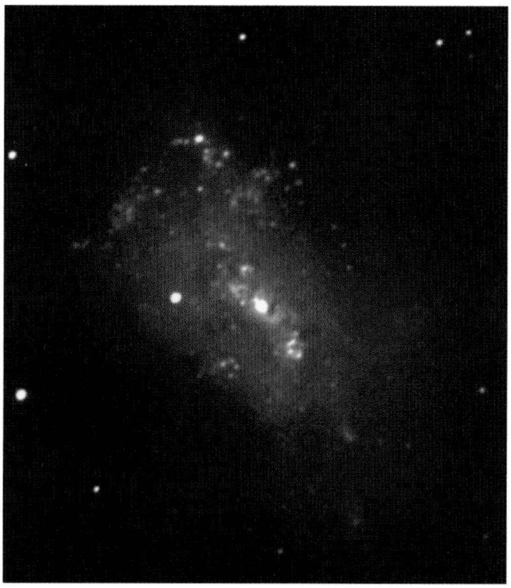

NGC 4449 imaged by Gordon Rogers of Long Crendon, Aylesbury, UK with a 0.4 m Schmidt-Cassegrain at *f*/6.3 plus SBIG ST8E CCD and AO7 adaptive optics unit. http://www.gordonrogers.co.uk

Caldwell 22

NGC 7662
Blue Snowball planetary nebula in Andromeda
Magnitude: 8.3
Size: 30″
Mag/Sec: 15
RA: 23 h 25 min 54 s
Dec: +42° 32′ 06″
Circumpolar: Above 48°N
Starhop: 13½°N and 8½°W from Alpheratz
Best Months: August to November N. hemisphere
Distance: ~3,000 light-years
Physical Size: ~5 light-months
Age: ~5,000 years
Best Visual Aperture: 15 cm and larger with high powers
Best Visual Filter: UHC or OIII
CCD/DSLR Advice: Tiny, so 1″/pixel and finer works best
Celestial Neighbors: Faint galaxy NGC 7640 is 2°S
Miscellaneous: Discovered by William Herschel in 1784

A number of the Caldwell objects we have encountered so far have been in Lacerta or, at least, not far from the Cygnus/Lacerta border. The Blue Snowball lives in western Andromeda not far from the opposite (eastern) border of Lacerta. Like most of the planetary nebulae in the Caldwell catalog NGC 7662 is small but of a high surface brightness. High magnifications are, once more, the best strategy, but finding the tiny planetary is quite a challenge. The nearest bright naked-eye star anywhere near Caldwell 22 is Alpheratz (α And), the top left star (from the northern hemisphere) in the square of Pegasus. A hop from there is possible but not really practical. The best strategy is arguably to find magnitude 3.6 o Andromeda (o = omicron), which lives near the Andromeda/Lacerta border at 23 h 02 min +42° 20′, center the star in the telescope, and then move 4½° due east. Once you suspect the 8th magnitude planetary, using magnifications of 400× and above on any telescope larger than 15-cm aperture will give the best results.

Despite this nebula's name it does not, to most viewers at least, look blue. Sorry about that! Virtually every faint object in the night sky looks either colorless or greenish to most. However, with large apertures it does, rumor has it, have a light blue cast to some eyes. It looks like a tiny elliptical gray-green ring with a fainter outer halo to most observers. The central star, as with many planetary nebulae's central stars, is a bit mysterious. Estimating the magnitude of a faint star against a dark background is straightforward, but when a nebulous backdrop is present the contrast is reduced and the nebula itself can even "dazzle" the fully dark-adapted and low-resolution rods. Images of Caldwell 22's central star indicate it is about magnitude 12.5, yet few – if any – visual observers have seen it.

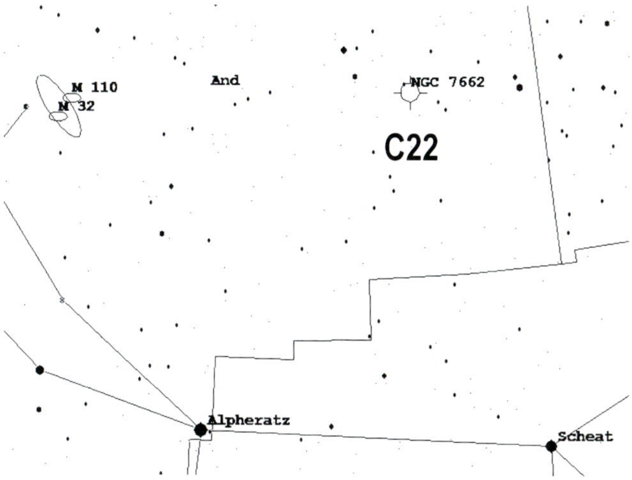

The chart shows the location of Caldwell 22 in Andromeda in relation to the northern edge of the Square of Pegasus and the Andromeda galaxy.

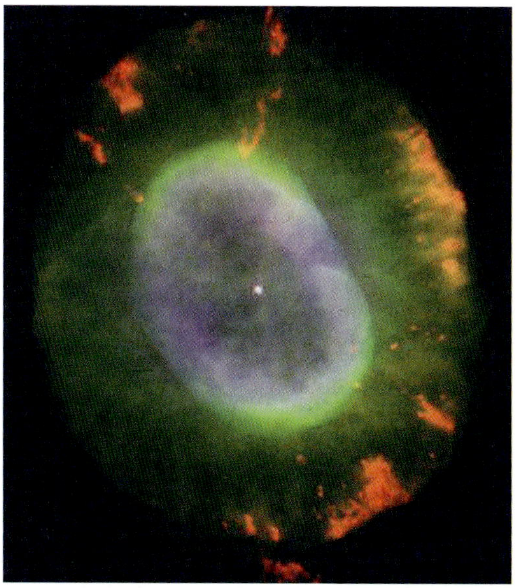

The Blue Snowball planetary nebula imaged at super high resolution with the Hubble Space Telescope. The image is only half an arc-minute wide, slightly less than the angular diameter of Jupiter as seen from Earth. Image: Howard Bond/NASA.

Caldwell 23

NGC 891
Edge on galaxy in Andromeda
Magnitude: 10
Size: 12′ × 3′
Mag/Sec: 23
RA: 02 h 22 min 36 s
Dec: +42° 21′
Circumpolar: Above 48°N
Starhop: 3½° due East of Almaak (γ Andromeda)
Best Months: September to January, N. hemisphere.
Distance: ~30 million light-years
Physical Size: ~105,000 light-years
Age: Billions of years
Best Visual Aperture: 20 cm and larger is recommended
Best Visual Filter: Deep sky filter may help
CCD/DSLR Advice: Superb with mono or LRGB imaging
Celestial Neighbors: Open cluster M 34 lies 4°E
Miscellaneous: Significant dark matter content

NGC 891 is one of the best known edge-on galaxies in the night sky, and it holds quite a few memories for this author. In the 1980s I spent some time fruitlessly searching for comets visually (300 h spread over 4 years, to be precise) with a 36 cm *f*/5 Newtonian. In that time I often stumbled over NGC 891. The only other galaxies I came across that looked so like a celestial splinter were NGC 5907 in Draco (actually called "The Splinter"), NGC 4565 in Coma Berenices (Caldwell 38), NGC 4244 ("The Silver Needle"/Caldwell 26) in Canes Venatici and NGC 4631 (Caldwell 32), also in Canes Venatici. I took my first photograph of NGC 891 on September 7, 1986, a 30-min exposure, guided visually with a guidescope atop that same 36 cm *f*/5 Newtonian and using 3M 1,000 film. A ten-second CCD exposure in 2009 easily blows that photo away! In images the galaxy resembles the proverbial (to quote Patrick) "two fried eggs clapped back to back." At more than 100,000 light-years in diameter NGC 891 is a big galaxy, similar in size to our own Milky Way. Tilt it face on to us, and it would be a prime supernova hunter's galaxy. Perhaps surprisingly a 14th magnitude supernova was discovered in this edge-on galaxy in the 1980s. Supernova 1986J was discovered on August 21 of that year by messrs. van Gorkom, Rupen, Knapp, and Gunn.

NGC 891 is very easy to find! It is conveniently situated 3½° due east of the bright second magnitude star Almaak (γ Andromedae), which anyone can find with ease. The galaxy is a very elusive object in any telescope under 10-cm aperture. Frankly, the bigger the better for Caldwell 23, and above 35-cm aperture you will see the central dust lane, with a bit of practice.

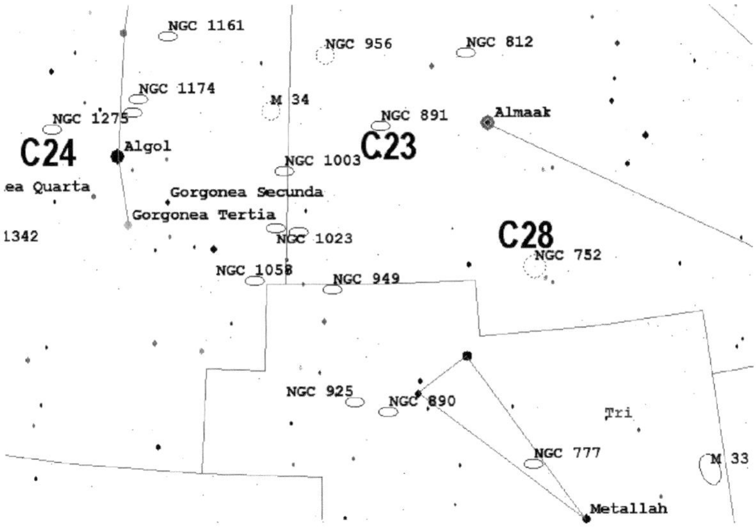

NGC 891 (C23) lies just east of the bright star Almaak as shown in the centre of the finder chart. C28, also in Andromeda and C24, in Perseus, are also shown.

The author captured this shot of NGC 891 on 2007 February 6 by coadding five 180 s exposures. Equipment: 0.35 m Celestron 14 at f/7.7 plus SBIG ST9XE CCD. Paramount ME mounting. Field 13 by 13 arcmin.

Caldwell 24

NGC 1275
Galaxy in Perseus
Magnitude: ~11.9 (nucleus variable)
Size: 3′ × 2′
Mag/Sec: 22
RA: 03 h 19 min 48 s
Dec: +41° 31′
Circumpolar: Above 49°N
Starhop: 2°E and ½°N from Algol (ß Persei)
Best Months: October to January, N. Hemisphere
Distance: 230 million light-years
Physical Size: ~200,000 light-years
Age: Billions of years
Best Visual Aperture: 30 cm and larger is recommended
Best Visual Filter: Deep sky filter may help
CCD/DSLR Advice: Try unfiltered imaging at long focal lengths
Celestial Neighbors: Faint galaxies everywhere
Miscellaneous: Brightest galaxy in massive Perseus Supercluster
Chart: See Caldwell 23

NGC 1275 seems a bit out of place in the Caldwell catalog. Of course, it is an extraordinary object from a professional astronomer's viewpoint – a strong radio emitting active galaxy (designated 3C 84 in the *Third Cambridge Catalog*) with a massive central black hole, often referred to simply as Perseus A. However, through any telescope of less than 20 cm aperture NGC 1275 is rather dismal, just a small, faint fuzz seen with averted vision, even if its surface brightness is quite reasonable. Through really large apertures the field comes alive with faint fuzzies as the galaxy is simply the brightest member of the Perseus cluster. What is really stunning is the knowledge that NGC 1275 lies at the mind-bending distance of 230 million light-years and so is the second furthest (see Caldwell 35) object in this catalog. When the light we see left Caldwell 24 it was the late Triassic period on Earth, and dinosaurs were just emerging. The location of Caldwell 24 is easy to find as it lies a mere 2° east and half a degree north of Algol in Perseus. In late 2007 this region was host to quite a spectacle when the exploding comet 17P/Holmes sailed past. Two supernovae have been found in NGC 1275. The first (1968A), discovered by Lovas in January 1968, reached 15th magnitude, although the type is not known; the most recent, 2005 mz, discovered by Jack Newton, Mike Peoples, and Tim Puckett peaked at 15.8 and was a Type Ia. Undoubtedly, NGC 1275 and its colleagues are more rewarding objects for the CCD imager than the visual observer. A deep 1° wide field will reveal a hundred galaxies to magnitude 17, even if many will look rather stellar in appearance. A monochrome image of the relatively colorless field looks just as impressive as an RGB or LRGB image.

This short exposure supernova patrol shot by Tom Boles, using a Celestron 14 at *f*/11 shows NGC 1275 and numerous other galaxies in Perseus in this 10 arcmin field. The bright galaxies near the top of the frame are NGC 1278 (brightest) and NGC 1277.

This dramatic Hubble Space Telescope image of NGC 1275 shows filamentary structures composed of cool gas suspended by a magnetic field. These filaments indicate energy is being transferred from a central massive black hole to the surrounding gas.

Caldwell 25

NGC 2419
Globular Cluster in Lynx
Magnitude: 10.4
Size: 4.6′
Mag/Sec: 22
RA: 07 h 38 min 09 s
Dec: +38° 52′ 55″
Circumpolar: Above 52°N
Starhop: 7°N and ½°E from Castor (α Gem)
Best Months: December to March, N. Hemisphere.
Distance: 280,000 light-years
Physical Size: ~370 light-years
Age: ~12 billion years
Best Visual Aperture: 15 cm and larger is recommended
Best Visual Filter: No filter advantage
CCD/DSLR Advice: Easy object in mono or color
Celestial Neighbors: Nothing in immediate vicinity
Miscellaneous: Furthest of the "bright" globular clusters

NGC 2419 is not the most awesome globular cluster visible in the night sky. In the northern hemisphere, M13 in Hercules shines almost 100 times brighter and is five times the diameter. In addition, where M13 has plenty of 11th magnitude stars that sparkle in a big 40 cm backyard telescope you will need to look through a 4 m instrument to see NGC 2419's brightest 16th magnitude stars. Let us not be too unkind to Caldwell 25, though. It is a long way away, at roughly 280,000 light-years. Put it side by side with M13, and it would be twice the size of that globular and even put the magnificent southern hemisphere globulars Omega Centauri and 47 Tucanae (Caldwell's 80 and 106) in their place. But you can still feel a sense of awe when you track NGC 2419 down; 280,000 light-years makes the globular an intergalactic freak, further out even than the Magellanic clouds. It is not quite a record breaker, as a few other globular clusters within the Milky Way's gravitational influence are even more distant, but it is still fascinating.

Although NGC 2419 is in Lynx it is close to the borders with eastern Auriga and northern Gemini, and the well known "twins" star Castor is the nearest naked eye pointer to the field. Simply finding Castor, nudging the telescope half a degree east, and then moving 7° north will get you to the field in those bone chilling northern hemisphere nights. No other tenth magnitude deep sky objects are anywhere near the field, but a distinctive couple of stars live nearby. Eight magnitude GSC 295827 and seventh magnitude GSC 2958330 are 5 arcmin apart (east-west), and if you continue the line another 5 arcmin east beyond the brighter star you arrive at the globular.

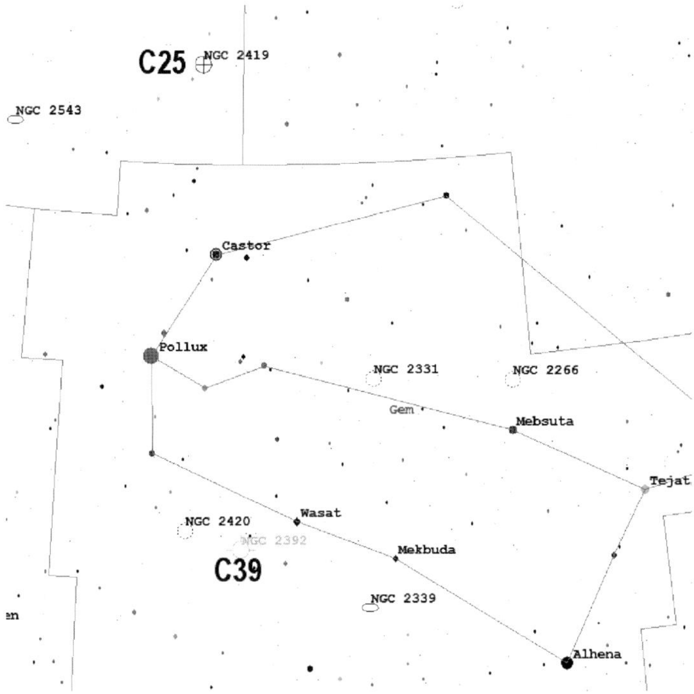

This finder chart shows the location of NGC 2419 (C25) in Lynx, not far north of Castor in Gemini. The position of Caldwell 39 in southern Gemini is also shown.

NGC 2419 from the Digitized Sky Survey. The field is 8 arcmin wide.

Caldwell 26

NGC 4244
Silver Needle Galaxy in Canes Venatici
Magnitude: 10.4
Size: 16′ × 2′
Mag/Sec: 23
RA: 12 h 17 min 30 s
Dec: +37° 48′
Circumpolar: Above 53°N
Starhop: ½°S and 7½°W from Cor Caroli (α Canes Venatici)
Best Months: March to June, N. Hemisphere
Distance: 10 million light-years
Physical Size: ~50,000 light-years
Age: Billions of years
Best Visual Aperture: 15 cm and larger is recommended
Best Visual Filter: Deep sky filter may help
CCD/DSLR Advice: Excellent object in mono or color
Celestial Neighbors: Galaxies NGC 4214 and 4151 within 2°
Miscellaneous: Discovered by William Herschel in 1787
Chart: See Caldwell 21

A few other edge-on deep sky "splinters" were mentioned when dealing with Caldwell 23, and here is one of those. This really is a full-blown splinter, as there is not a hint of the central bulge so obvious in that previous object. NGC 4244 is longer, thinner, and slightly fainter than NGC 891 and really does look plain weird through the eyepiece of a big telescope. By virtue of its weird appearance alone Caldwell 26 is worth chasing down. The galaxy lives not far from another Caldwell galaxy we mentioned a few pages back, number 21, or NGC 4449. Once again, locating the bright star Cor Caroli is the first step. From there move half a degree south and 7½° west. No other bright galaxy is closer than a degree and a half of the silver needle that is NGC 4244. The name, apparently, was coined by the deep sky enthusiast Tom Polakis. In small telescopes, the galaxy is quite a challenge, although its unusual shape triggers a subconscious "what the heck is that" response when it drifts across the retina's rods. In 30-cm apertures and larger the low surface brightness of the galaxy becomes less of an issue, and Caldwell 26 can be held steady. The center of the silver needle appears obviously brighter than the ends in large apertures and some texture is apparent. The impressive length of NGC 4244 makes it a prime target for Schmidt-Cassegrain imagers with fields of view in the quarter-degree category. The galaxy has conveniently angled itself to lie northeast-southwest, enabling the CCD chip diagonal to fully exploited. Caldwell's 25–29 all lie between +39° and +37° and the position of Caldwell 26 means that, strictly speaking (like the Hyades/C41), it is two Caldwell numbers too low. It actually belongs just after the existing Caldwell 28, but we are not going to change the sequence!

The edge-on galaxy NGC 4244 in Canes Venatici, imaged by the supernova discoverer Ron Arbour from South Wonston in the UK. A fork mounted 0.3 m Schmidt-Cassegrain working at *f*/5 with a Starlight Xpress SXV-H9 CCD was used.

Caldwell 27

NGC 6888
Crescent Nebula in Cygnus
Magnitude: 8.8
Size: 2′ wide, 20′ long crescent
Mag/Sec: 22
RA: 20 h 12 min 00 s
Dec: +38° 21′
Circumpolar: Above 52°N
Starhop: 2°S and 2°W from Sadir/Sadr
Best Months: June to October, N. Hemisphere.
Distance: 4,700 light-years
Physical Size: ~25 light-years
Age: ~100,000 years
Best Visual Aperture: 15 cm and larger is recommended
Best Visual Filter: OIII or UHC
CCD/DSLR Advice: Crescent enhanced by H-Alpha luminance
Celestial Neighbors: M29, IC 4996, NGC 6883 and 6871
Miscellaneous: Precise nature is uncertain
Chart: See Caldwell 15

Cygnus, the Swan, is one of the most easily identified northern hemisphere constellations, although to most observers it simply looks like a straightforward cross and nothing like a flying bird. At the center of the cross you will find the second magnitude star Sadir, and the Crescent Nebula is just a short hop 2° south and 2° west from there. Another showpiece, the open cluster M29 is nearby, just a degree and a half and slightly east of that same star, Sadir. The Crescent is a great object to go for on those warm August and September evenings in the northern latitudes, arguably the only 2 months of the year there when dark skies can be enjoyed without the risk of hypothermia. The field seems to be around forever, too, as the westward drift of Cygnus in evening twilight seems to be exactly counterbalanced by the northern evenings drawing in after August. As soon as it gets dark Cygnus seems to be overhead month after month. The brightest star within the curve of the Crescent is magnitude 7.5 V1770 Cyg, also known as GSC 3151 1765. This is a massive Wolf-Rayet star with a binary component. Wolf-Rayet stars are well known for generating strong interstellar winds, and the Crescent may be the result of emissions from a relatively recent outburst striking older remnants of the star's atmosphere. Images from IRAS (the Infra-Red Astronomy Satellite) have shown another nebula shell outside the crescent shape, which could be the result of an old supernova explosion. However, many astronomers think it just an earlier ejection from the same Wolf-Rayet star. Therefore, the Crescent Nebula may or may not be associated with a supernova remnant, but it is still a fascinating object to track down, visually or photographically. The use of a nebula filter will help considerably.

The Crescent Nebula, NGC 6888, imaged by UK astro-imager Ian Sharp of Ham, West Sussex. An 80 mm semi-apochromatic Orion ED80 refractor (600 mm focal length, *f*/7.5) mounted on an Astrophysics AP900 GTO was used. The camera was an ATIK 314L. The image consists of 21 ten minutes exposures through a Hydrogen alpha filter and 18 ten minute exposures through an OIII filter. The OIII data was used for the B component and a synthetic G was created by merging H-alpha and OIII data. H-alpha was used for R and for the luminance data (50% opacity setting).

Caldwell 28

NGC 752
Open Cluster in Andromeda
Magnitude: 5.7
Size: 50′
Mag/Sec: Not relevant
RA: 01 h 57 min 36 s
Dec: +37° 50′
Circumpolar: Above 53°N
Starhop: 4½°S and 1°W from Almaak (γ Andromeda)
Best Months: September to January, N. Hemisphere.
Distance: ~1,200 light-years
Physical Size: ~20 light-years
Age: ~2 billion years
Best Visual Aperture: Big binoculars are good enough
Best Visual Filter: No filter required
CCD/DSLR Advice: Bland object, best with wide fields (>1°)
Celestial Neighbors: C 23 (NGC 891) 6° to the NE
Miscellaneous: Ancient, except compared to Caldwell 1!
Chart: See Caldwell 23

For the 28th Caldwell object we are back to the vicinity of Almaak (γ Andromeda), where we last journeyed for Caldwell 23. But this time, instead of going due east, we head 4½° south and 1° west. Andromeda is a pretty barren constellation and perhaps not the sort of place you might expect to find an open cluster; after all, the Milky Way's edge is almost 10° further north near the border with Perseus. Mind you, Caldwell 28 is probably not the sort of open cluster you might be familiar with; it is very big, roughly 50 arcmin across, and very loose. Low powers are a distinct advantage with this cluster, as at high powers you lose the edge and NGC 752 simply resembles a rich starfield. A pair of 10 × 50 binoculars will give you a pleasing view. Two distinctive magnitude 6 stars flank the field half a degree south-southwest of the cluster center, lying just outside the visual boundary of the system. Within the cluster diameter itself the brightest star is magnitude 7 (GSC 2816111) and forms an obvious tiny triangle with two fainter (magnitude 9) stars to the south. The majority of the cluster's stars are magnitude 10 or 11, and there are several dozen of them. The problem is, although the cluster itself is best viewed as a fuzzy cluster with low power, small instruments, you need something rather bigger to bring these stars out into individual points. If you sweep the field with those tiny aperture sports binoculars that seem to have become popular in recent years (typically 10 × 25 mm) it is interesting to note how similar in size and brightness the famous pinwheel galaxy M33, some 7° to the south-southwest, appears. Both objects are close to magnitude 6, and both look a bit less than a degree across.

The central region of NGC 752 imaged by Robert Gendler. Images taken with a 12.5-in. *f*/9 (318 mm) Ritchey-Chrétien and SBIG STL11000 CCD. LRGB = 15:15:15:15 min. © Robert Gendler.

Caldwell 29

NGC 5005
Galaxy in Canes Venatici
Magnitude: 9.8
Size: 6′ × 3′
Mag/Sec: 21
RA: 13 h 10 min 54 s
Dec: +37° 03′
Circumpolar: Above 53°N
Starhop: 1¼°S and 3°E from Cor Caroli (α Canes Venatici)
Best Months: March to June, N. Hemisphere
Distance: ~70 million light-years
Physical Size: ~120,000 light-years
Age: Billions of years
Best Visual Aperture: 15 cm and larger
Best Visual Filter: Deep sky filter may help
CCD/DSLR Advice: Pretty in deep LRGB images
Celestial Neighbors: M63 and M94 within 5°
Miscellaneous: Low energy Seyfert-type galaxy
Chart: See Caldwell 21

We are back in the northern spring galaxy fields for Caldwell 29 and a modest looking galaxy in Canes Venatici, the third of the seven Caldwell galaxies not far from the Canes Venatici/Coma Berenices border. Yet again, the star Cor Caroli can be used as a starting point. From there simply hop 1¼° south and 3° east. In any instrument larger than 10 cm a pleasing ellipse with a bright center will be seen. In many amateur CCD images of NGC 5005, the galaxy looks less than impressive, namely, a rather bland oval with a bright central ellipse and much fainter outer regions. However, in the deepest images Caldwell 29 is a very pretty catch, even if the outer spiral arms are rather faint and subtle objects. An LRGB image with a very long cumulative luminance component is necessary to bring out those arms which, even in the best amateur images, look browny-gray, not blue. In a small telescope, it is easy to get confused by another slightly bigger galaxy of the same magnitude, just half a degree to the southeast, namely NGC 5033. Frankly, both galaxies are equally deserving of a Caldwell status. NGC 5033 is, in effect, Caldwell 29B.

One supernova has been discovered in NGC 5005, by Claudio Bottari on June 16, 1996. The find was announced on IAU circular 6422. Supernova 1996ai brightened from magnitude 14.5 to magnitude 13 and was a highly prized Type Ia. However, NGC 5033 has been far more productive, with three supernovae being discovered in that galaxy in 1950 (type unknown), 1985 (type II), and 2001 (type IIb), the latter object found by the prolific Japanese amateur supernova hunter Itagaki).

NGC 5005 imaged by Robert Gendler. Images taken with a 12.5-in. *f*/9 (318 mm) Ritchey-Chrétien and SBIG STL11000 CCD. LRGB = 140:20:20:40 min. © Robert Gendler.

Caldwell 30

NGC 7331
Galaxy in Pegasus
Magnitude: 9.5
Size: 10′ × 5′
Mag/Sec: 22
RA: 22 h 37 min 06 s
Dec: +34° 25′
Circumpolar: Above 56°N
Starhop: 6½°N and 6°W from Scheat (ß Peg)
Best Months: August to November, N. Hemisphere.
Distance: 50 million light-years
Physical Size: ~150,000 light-years
Age: Billions of years
Best Visual Aperture: 15 cm and larger
Best Visual Filter: Deep sky filter may help
CCD/DSLR Advice: Pretty in deep LRGB images
Celestial Neighbors: Stephan's Quintet (see text)
Miscellaneous: 50 years since its last supernova

NGC 7331 is one of the best known galaxies in the northern celestial hemisphere not in the Messier catalog. Apart from its healthy ninth magnitude brightness and 10 arcmin length, there is some superb structure visible in the best amateur images. At any shallower angle to us this big galaxy would lose its appeal, but fortunately, at almost 70° from being face-on there is still plenty of detail. In fact, the foreshortening effect arguably gives it a more three-dimensional appearance. In color images an almost golden center contrasts perfectly with the blue arms: superb! NGC 7331 is not alone either. In a 15 arcmin-wide CCD field a retinue of four smaller 14th and 15th magnitude galaxies to the east can be squeezed into the picture. These faint fuzzies are NGC's 7335, 7336, 7337, and 7340. In 1959 Milton Humason discovered a bright (13th mag, Type II) supernova 32 arcsec west and 13 arcsec north of the galaxy center. Visually, Caldwell 30 is a rewarding object, too. Even for a predominantly digital imager in a 25 cm Newtonian the center is highly condensed and the oval shape very obvious. In larger telescopes the spiral arms are easy to spot.

Not far from the field, namely, half a degree to the south-southwest, another famous northern Pegasus deep sky imaging target can be found. Stephan's Quintet is, not surprisingly, composed of five galaxies (NGC's 7317, 7318A, 7318B, 7319, 7320) that are all either 13th or 14th magnitude. They easily fit within a 4-arcmin field. While the last and largest (in angular terms) of these galaxies appears to be at a similar distance to NGC 7331 the other four appear to be eight times more distant, at least, if redshift is always an indicator of distance, which not everyone believes.

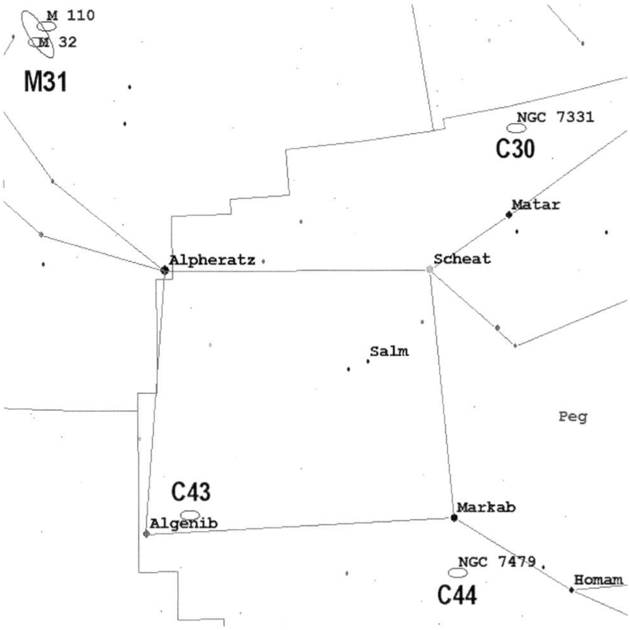

This chart is dominated by the regions of Pegasus and Andromeda. NGC 7331 (C30) lies to the northwest of Scheat, the top right star in the Square of Pegasus. Caldwell's 43 and 44 also reside in Pegasus and the Andromeda galaxy M31 is shown in the *top left*.

NGC 7331 imaged on October 22nd 2008 by the author from Cockfield, Suffolk, U.K. using a 35 cm Celestron 14 Schmidt-Cassegrain at *f*/7.7 and an SBIG ST-9XE CCD camera. Thirteen 120 s unguided images were stacked. Twelve arc-minute field. Paramount ME mounting.

Caldwell 31

IC 405
Flaming Star Nebula in Auriga
Magnitude: ~9
Size: 50′ × 30′
Mag/Sec: 24 (brightest region)
RA: 05 h 16 min 30 s
Dec: +34° 20′
Circumpolar: Above 56°N
Starhop: 5¾°N and 2°W from El Nath (ß Tau)
Best Months: November to February, N. Hemisphere
Distance: ~1,500 light-years
Physical Size: ~20 light-years
Age: AE Aurigae ejected 2.5 million years ago (see below)
Best Visual Aperture: Try 20 cm aperture at ~40×
Best Visual Filter: Deep sky or UHC filter may help
CCD/DSLR Advice: Wide H-alpha shots show the most detail
Celestial Neighbors: Five clusters to the east, inc. M 36 and M 38
Miscellaneous: AE Aurigae zips along at more than 100 km/s

Southern Auriga will be familiar territory for many northern hemisphere deep sky observers, but probably for better known objects than IC 405. The region lies dead in the middle of the winter Milky Way, and the open clusters M36 and M38 are less than 5° to the east and northeast, respectively. The first magnitude star El Nath (ß Tau), bizarrely shared by Taurus and Auriga, lies only 6° to the south-southeast.

IC 405 is not an exciting object visually, which is probably why it is rarely mentioned in amateur astronomy magazines and books. However, when you know the cause of the diffuse nebula it becomes a lot more interesting. The nebula is illuminated by the bright variable star AE Aurigae, which can easily be studied with binoculars as it varies between magnitude 5.8 and 6.1; not a huge variation, but enough for visual observers to monitor. It turns out that AE Aurigae is not a natural resident of IC 405; it is actually ploughing through the nebula at a speed of about half a million kilometers per hour. At that speed, roughly 10 times as fast as any Earth-launched spacecraft yet built, the Earth–Moon distance would be traversed in under an hour, and a trip from Earth to Saturn would take a few months.

What is the cause of AE Aurigae's rapid motion? Well, astronomers think it was ejected from the Orion Nebula's Trapezium system, some 1,300 light-years from IC 405, about 2.5 million years ago, along with another fast moving star, μ Columbae. The cause of the ejection may well have been the near-merger of two binary star system, with stars gravitationally swapping partners and the old partners being hurled into oblivion in a sort of interstellar divorce! Although the nebula illuminated by AE Aurigae is 50′ × 30′ wide in images, visually it is, at best, half that size. It is best seen at low powers such as 40× in a 20-cm telescope.

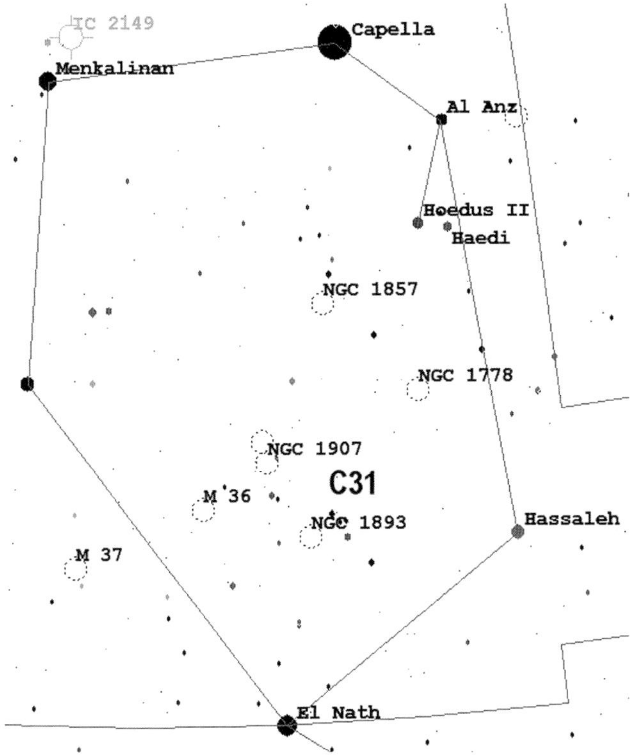

The Flaming Star Nebula IC405 (C31) lies in southern Auriga as shown in the chart.

IC 405 imaged by Tom Harrison using a 10-in. (254 mm) Schmidt-Cassegrain at *f*/5.75 with an H-alpha filter and SBIG ST-10XME CCD with AO-7 adaptive optics unit providing the guiding. The exposure was 10 min. AE Aurigae is the brightest star in the picture, to the *bottom left* of centre.

Caldwell 32

NGC 4631
The Whale Galaxy in Canes Venatici
Magnitude: 9.2
Size: 14.8′ × 3.5′
Mag/Sec: 22
RA: 12 h 42 min 06 s
Dec: +32° 32′
Circumpolar: Above 58°N
Starhop: 6°S and 3°W from Cor Caroli (α Canes Venatici)
Best Months: March to June, N. Hemisphere
Distance: 22 million light-years
Physical Size: ~95,000 light-years
Age: Billions of years
Best Visual Aperture: 15 cm and larger
Best Visual Filter: Deep sky filter may help
CCD/DSLR Advice: Easy target with a ¼° field
Celestial Neighbors: NGC 4656 ½° to SE
Miscellaneous: A bruised and battered galaxy
Chart: See Caldwell 21

Guess what? We are back near Cor Caroli yet again, with that same old group of seven galaxies. NGC 4631 is yet another surprisingly bright edge-on galaxy not cataloged by Charles Messier. In terms of size and brightness it is not dissimilar to a stretched version of M 82. Move 6° south and 3° west of Cor Caroli and you will spot this streak of a galaxy in any modest telescope. However, just as with Caldwell 29, there is another noteworthy galaxy not far away in the form of tenth magnitude NGC 4656, half a degree to the southeast. In CCD images each bright galaxy has a much fainter companion. In the case of NGC 4631, the tiny, 13th magnitude NGC 4627 lies 2 arcminutes to the north and appears to be joined to the larger galaxy, whereas for NGC 4656, the tiny, also 13th magnitude NGC 4657 chops across the former galaxy's northeast flank, giving rise to that pair's moniker of "The Hockey Stick."

Anyway, what does NGC 4631 actually look like through a telescope? Well, it is quite impressive in a big scope with plenty of subtle texture and mottling. Although the impressive features in this object are well known and admired by deep sky enthusiasts, outside the keenest circles NGC 4631 seems to almost be a well-kept secret! Perhaps a non-Messier edge-on galaxy is just not expected to look this detailed? There are two reasons for this wealth of detail. First, the galaxy is a tortured one, stretched by its tiny companion and NGC 4656/7. Second, it has a very respectable surface brightness, making these subtleties easy to see. In deep images the detail can be stunning: a vivid blue disc with a subtle yellow bulge in the background. With a vivid imagination the disc of the starship *Enterprise* can be seen heading towards us, deck lights ablaze!

NGC 4631 imaged by Gordon Rogers of Long Crendon, Aylesbury, UK with a 0.4 m Schmidt-Cassegrain at *f*/6.3 plus SBIG ST8E CCD and AO7 adaptive optics unit. http://www.gordonrogers.co.uk

Caldwell 33 and Caldwell 34

NGC 6992/6995 & 6960
Network, Veil, and Filamentary nebulae – Supernova remnants in Cygnus
Magnitude: both main arcs roughly magnitude 8
Size: Two arcs lying on a 1¼° radius
Mag/Sec: ~22–24
Positions for the 3 brightest sections:

C33/NGC 6992	RA: 20 h 56 min 30 s
	Dec: +31° 40′
C33/NGC 6995	RA: 20 h 57 min 05 s
	Dec: +31° 13′
C34/NGC 6960	RA: 20 h 45 min 40 s
	Dec: +30° 45′

Circumpolar: above 60°N
Starhop: 3¼°S from Gienah (ε Cyg) takes you to 52 Cygni/NGC 6960, the W. edge
Best Months: July to October, N. Hemisphere.
Distance: ~1,500 light-years
Physical Size: ~70 light-years
Age: ~5,000–8,000 years (see text)
Best Visual Aperture: 10 cm and larger with dark skies and filters
Best Visual Filter: UHC filter works best
CCD/DSLR Advice: Long H-alpha exposures work best
Celestial Neighbors: Entire neighborhood rich in nebulosity
Miscellaneous: Exceptional supernova remnant
Chart: See Caldwell 15 for a wider chart of the region

Included here together are the two Caldwell objects involved in this feature, as it would be pointless to deal with them separately. The object concerned is called the Veil (sometimes called the Bridal Veil) Nebula in Cygnus. However, the eastern part can be called the Network Nebula and the western part is sometimes referred to as the Filamentary Nebula.

If you go back prior to the 1970s, objects as ghostly as the Veil were almost considered "impossible" targets for the visual observer. The low surface brightness certainly makes it a challenging object, but the availability of high contrast and emission line nebula filters largely pioneered by companies such as Lumicon have helped pull the ghostly details out of the skyglow. There are a number of good deep sky filters on the market, but the best ones for SNR viewing are those that pass the two Oxygen III emission lines at 496 and 501 nm, such as the Lumicon Oxygen III filter, the Celestron Oxygen III Narrowband filter, and the Meade Series 4000 Oxygen III filter. Ultra high contrast (UHC) filters that pass the O III lines and the hydrogen beta line at 486 nm are excellent, too. Indeed, Stewart Moore, director of the BAA Deep Sky Section, vastly prefers the enhancing effect of a general UHC filter on the Veil, regarding an OIII filter as disappointing by comparison. However, those living in fairly light-polluted sites will prefer the narrower O III filters, as they darken the background more and there are only a few nebulae that benefit from a hydrogen beta window. All other objects, apart from O III emitters, will be severely dimmed by an O III filter, but the contrast enhancement on some supernova remnants can be stunning. It is amazing that such targets were considered

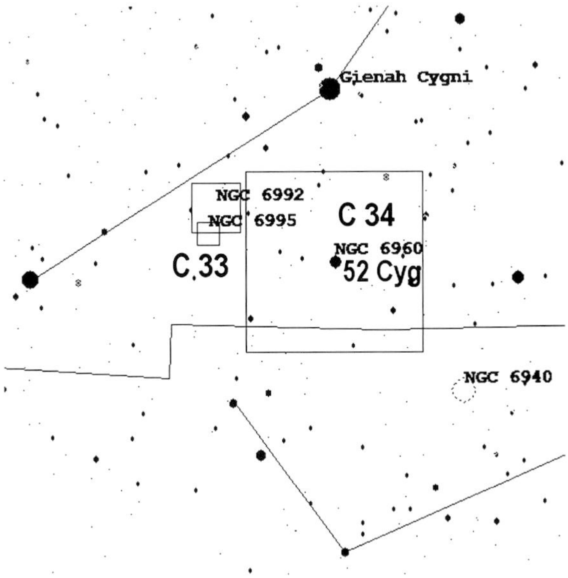

A 10° wide field showing the relative positions of Caldwell's 33, 34, the mag. 2 star Gienah Cyg. and the mag. 4 star 52 Cyg.

NGC 6992 (*upper*) and NGC 6995 (*lower*), the Eastern Veil, imaged by Ian Sharp. This is a two part mosaic of this large object compiled by the UK astro-imager Ian Sharp of Ham, West Sussex. An 80-mm semi-apochromatic Orion ED80 refractor (600 mm focal length, *f*/7.5) mounted on an Astrophysics AP900 GTO was used. The camera was an ATIK 314L. Each half of the mosaic consists of a stack of 18 ten minute exposures through an H-alpha filter.

virtually impossible to see prior to the 1970s. Another factor here is the increasing willingness of amateurs to transport powerful instruments to very dark sites and to indulge in communal events such as star parties. At such events there are always experienced observers, large apertures, and a range of nebula filters for all to see and use.

However, before we get carried away with aperture fever, it is important to stress that you do NOT need a massive telescope to see the Veil Nebula. It is, perhaps surprisingly, a very large deep sky object whose two halves are part of a circle almost 3° in diameter. This is not dissimilar in size to the famous Andromeda Galaxy, M31. However, it is the surface brightness of the two arcs that is the challenge and not its size, hence the importance of dark skies and appropriate filters. The Veil Nebula's two halves are in a very familiar part of the summer sky for northern hemisphere observers. They are situated just a few degrees south of the easternmost star of Cygnus's famous five star cross, namely epsilon, or Gienah Cygni. Gienah is a magnitude 2.5 easy naked-eye star and is situated at a declination of +34°. Move just 3° or so south of Gienah and you will find the 4th magnitude star 52 Cygni. The Veil's western component, NGC 6960, actually passes through the field of this star, which is both a curse and a blessing. It makes the western component easy to locate, but to prevent the star from dazzling your view you will be tempted to nudge the telescope so that 52 Cygni is outside the field. In a quality, wide-field telescope used at 20× you will just be able to fit the whole Veil in your field of view. With a wide field of just more than 2.5° and 52 Cygni at the western edge of the field, the eastern section just fits in. The western arc has the designation NGC 6960, and the eastern arc is composed of NGC 6992 and NGC 6995. The Veil arc is not complete. There is a major chunk missing to the southeast, even in deep photographs and images, but it is a beautiful and delicate sight in a dark sky, especially if you have an Ultra High Contrast filter.

The current thinking is that the Veil Nebula actually lies at a distance of roughly 1,500 light-years from Earth, based on observations made by the Hubble Space Telescope in 1999. (As an aside, this is more than seven times further than 52 Cygni, which is about 200 light-years from Earth). This 1,500 light year distance makes the Veil five times closer than the Crab Nebula, M1. In terms of its age in our night sky though, the Veil is much older. The supernova responsible for its formation probably exploded in our skies around 5,000–8,000 years ago. However, it is not inconceivable that both the Crab and the Veil supernovae actually went bang at about the same time. Ignoring, for simplicity sake, the fact that since Einstein's relativity theories, there has been no such thing as an absolute time clock, it is possible that both the Veil and the Crab supernovae detonated, say, 8,000 years ago. But the light from the Veil supernova would have arrived after 1,500 years and that from the Crab after 7,000 years. In a large galaxy, the finite speed of light can sometimes look very slow indeed and completely skews our perception of events. If the Veil supernova was only 1,500 light-years away, it may well have been a dazzling magnitude-9 object in the skies of our ancestors.

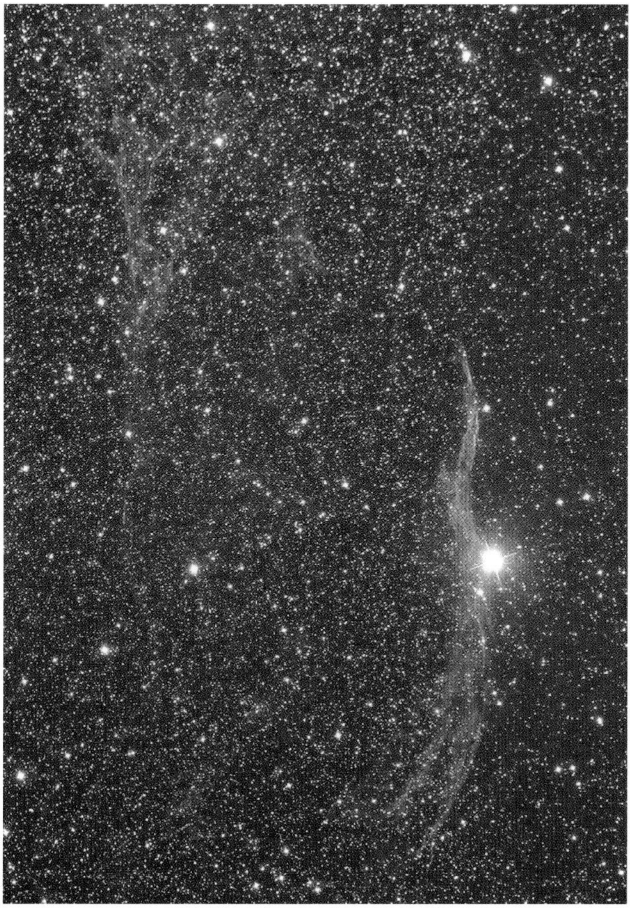

The western half of the Veil nebula, NGC 6960 and the bright star Kappa (52) Cygni which sits on the extreme western end of the Veil. A color stack of sixteen 180 s images by Ian Sharp with a Takahashi Epsilon 130 mm Astrograph (*f*/3.6) and Hutech-modified Canon 350D on an Astrophysics AP1200 GTO mount.

Caldwell 35

NGC 4889
Galaxy in Coma Berenices
Magnitude: 11.5
Size: 3′ × 2′
Mag/Sec: 22
RA: 13 h 00 min 08 s
Dec: +27° 59′
Circumpolar: Above 63°
Starhop: 1°E and 10½°S from Cor Caroli (α Canes Venatici)
Best Months: March to June, N. Hemisphere.
Distance: ~300 million light-years
Physical Size: ~250,000 light-years
Age: Billions of years
Best Visual Aperture: 30 cm and larger
Best Visual Filter: Deep sky filter may help
CCD/DSLR Advice: Try mono imaging at long focal lengths
Celestial Neighbors: Many faint galaxies nearby
Miscellaneous: Most distant Caldwell object
Chart: See Caldwell 21

Eleven Caldwell objects ago we encountered Perseus A (NGC 1275), a rather dim object but the brightest galaxy at the center of a swarm of very distant galaxies. Well, here we go again, except NGC 4889 is even more distant. Whereas a distance of 230 million light-years has been allocated to Perseus A, an even more mind-bending distance of 300 million light-years is the best guess for NGC 4889. Unless the red shift based cosmological distance scale is in error (as Halton Arp has suggested for the last twenty years), NGC 4889 is the most distant Caldwell object. However, it is certainly not the most distant object visible in an amateur telescope, as a number of quasars billions of light-years away are within amateur reach. Another consequence of NGC 4889's distance is that, with a respectable 3 arcmin length, the galaxy must be a quarter of a million light-years across – a real celestial giant. By comparison our own Milky Way Galaxy is thought to be 100,000 light-years across, and the large Andromeda Galaxy, M31, is thought to be 140,000 light-years in diameter. So NGC 4889 is a colossus!

Caldwell 35 can be found just 3° south of the northern Coma Berenices border. Cross that boundary and, once again, we end up in southern Canes Venatici, such a popular home for Caldwell galaxies! Although our old favorite Cor Caroli can be used as a pointer to the field (just head 10½° south and 1° east) finding fourth magnitude ß Comae (13 h 12 min and almost +28° Dec.) and sliding almost 3° west will get you there, too. A trio of stars (two of magnitude 7 with a magnitude 8. vertex) lie just northwest of the field, with the star GSC 19951732 being closest. Through a telescope the galaxy is just a smudge, but a half degree wide CCD image will reveal a dozen fuzzy companions.

NGC 4889 imaged by Adam Block using a 16-in. RC Optical Systems telescope Operating at *f*/8 at the Kitt Peak visitors centre. Paramount ME Robotic Telescope Mount. SBIG ST10XME CCD camera with color filter wheel. LRGB: 180:20:10:20 min. RGB frames binned 2 × 2. *CCDsharp* deconvolution and Digital Development (DDP) were used to process the image. Image credit: Adam Block/NOAO/AURA/NSF

Caldwell 36

NGC 4559
Galaxy in Coma Berenices
Magnitude: 10.0
Size: 11′ × 5′
Mag/Sec: 23
RA: 12 h 36 min 00 s
Dec: +27° 58′
Circumpolar: Above 63°N
Starhop: 10½°S and 4°W from Cor Caroli (α Canes Venatici)
Best Months: March to June, N. Hemisphere.
Distance: 32 million light-years
Physical Size: ~100,000 light-years
Age: Billions of years
Best Visual Aperture: 20 cm and larger
Best Visual Filter: Deep sky filter may help
CCD/DSLR Advice: Easy unfiltered or LRGB target
Celestial Neighbors: Many faint galaxies nearby
Miscellaneous: ~70 years since its bright 1941 supernova
Chart: See Caldwell 21

We do not have to nudge the telescope far to travel from Caldwell 35 to Caldwell 36; just pushing it almost 5½° due west will get you to NGC 4559. This galaxy looks considerably bigger in the eyepiece, and for a good reason: it is almost 10 times closer. While a long starhop from Cor Caroli is possible, identifying fourth magnitude γ Comae (12 h 27 min, +28° 16′) and hopping 2° east and a quarter of a degree south is easier once you have spotted that star. The field is only fractionally outside (northeast of) the 4° circle that is the star cluster Melotte 111, easily visible in small binoculars. In a 15 cm telescope, Caldwell 36 is immediately obvious as a distinct oval smudge; roughly half its true length is revealed to the eye in a brief study. A hint of spiral structure can be held steady in larger instruments.

In long exposure, high-resolution CCD images NGC 4559 is revealed as another stunner, with the acute viewing angle giving depth to the galaxy. Color images show silvery-blue inner spiral arms sprinkled with pinkish H-II regions and a more subtle, fainter outer halo filling out the remaining diameter. Bright star-forming regions and dust lanes are clearly visible. We will shortly come to another stunning galaxy 2° further south in the form of Caldwell 38. On February 5, 1941, a magnitude 13.2 supernova was discovered 30 arcsec west and 26 arcsec north of the center of NGC 4559 by Jones. The supernova was designated 1941A, as it was the first supernova discovered in that year, in the middle of World War II. Only two other supernovae were discovered in 1941 and only ten during the five-year duration of the war.

NGC 4559 imaged by Gordon Rogers of Long Crendon, Aylesbury, UK with a 0.4 m Schmidt-Cassegrain at *f*/6.3 plus SBIG ST8E CCD and AO7 adaptive optics unit. http://www.gordonrogers.co.uk

Caldwell 37

NGC 6885
Open Cluster in Vulpecula
Magnitude: 8.1
Size: 18′
Mag/Sec: Not relevant
RA: 20 h 11 min 36 s
Dec: +26° 28′
Circumpolar: Above 64°N
Starhop: 1½°S and 9°E from Albireo (ß Cyg).
Best Months: June to October
Distance: 1,900 light-years
Physical Size: ~10 light-years
Age: ~15 million years
Best Visual Aperture: Big binoculars and larger
Best Visual Filter: No filter required
CCD/DSLR Advice: Easy target
Celestial Neighbors: Dumbell Nebula M27, 4° to SSW
Miscellaneous: Include it with the Coathanger
Chart: See Caldwell 15

At 26° in declination we have now reached parts of the celestial sphere that are reasonably accessible from both hemispheres of Earth, excepting the really far southern locations, of course. This has brought us, for the first time, into Vulpecula, a constellation perhaps best known for its prolific production of binocular brightness novae, despite its modest size. The identity of the open cluster NGC 6885 is somewhat confused by the fact that it seems, in some modern catalogs, to be the same, or maybe not quite the same, as NGC 6882! Some sources seem to indicate there are two clusters at NGC 6885's position; sometimes one cluster is superimposed on the other, and sometimes they are separate. Frankly, it is such a historical muddle that we will not bore you here with the tale. Essentially, if you search out the magnitude 6 star 20 Vulpeculae and view the region at low powers (with a half degree field) you will see a lot of stars in a somewhat loose congregation around the region. Quite whether you can see one cluster or two is up to you, but the presence of 20 Vul in the field complicates the brain's interpretation of the scene. The field contains a healthy smattering of ninth and tenth magnitude stars and loads more down to 13th magnitude. Locating 20 Vul can be achieved by moving 9° east and a degree and a half south from the glorious double star Albireo. In low power binoculars with a 5° field, just moving two fields to the left (in the northern hemisphere) should bag it; 20 Vul is less than 2° southwest of fourth magnitude 23 Vul, and half a degree to 20 Vul's north/northwest the fifth magnitude 19 and 18 Vul act as pointers to the field. Honestly, when you are near this region you might just prefer to look at the Coathanger (Collinder 399 – 19 h 25 min +20° 11′).

NGC 6885 from the Digitized Sky Survey. The field is 15 arcmin wide.

Caldwell 38

NGC 4565
Galaxy in Coma Berenices
Magnitude: 9.6
Size: 16′ × 2.5′
Mag/Sec: 22
RA: 12 h 36 min 18 s
Dec: +25° 59′
Circumpolar: Above 65°N
Starhop: 11½°N and 11°E from Denebola (ß Leo)
Best Months: February to May
Distance: 30 million light-years
Physical Size: ~140,000 light-years
Age: Billions of years
Best Visual Aperture: 15 cm and larger
Best Visual Filter: Deep sky filter may help
CCD/DSLR Advice: Superb galaxy unfiltered or in color
Celestial Neighbors: Many faint galaxies nearby
Miscellaneous: One of the eeriest sights in the sky
Chart: See Caldwell 21

Mentioned earlier was the magnificent galaxy NGC 4565 in the context of the northern skies' other best edge-on specimens, NGC's 891, 4244, 4631, and 5907. The former three, if you need reminding, are Caldwell's 23, 26, and 32. (NGC 5907 is not in the Caldwell catalog). NGC 4631 (the Whale) is not quite edge-on, but arguing over a 5° tilt is splitting hairs. All these galaxies are similar but subtly different. In a contest NGC 4565 would probably win as the definitive edge-on galaxy with all the attributes required. It is far more like Patrick's "two fried eggs clapped back to back" than NGC 891 and is brighter and longer, too. NGC's 4244 and 5907 are called the "Silver Needle" and the "Splinter" for a reason; while attractive, the lack of a genuine fried egg central bulge lessens their beauty. As for NGC 4631, well, as we have seen, it is an absolute stunner but cannot match NGC 4565 if you are looking for a definitive edge-on galaxy.

To find Caldwell 38 the best strategy is to view it immediately after Caldwell 36. These two galaxies are the sixth and seventh Patrick chose in that favorite Canes Venatici/Coma Berenices border region of his. Moving 11½° north and 11° east from Denebola in Leo gets you there, but so does moving 2° south from Caldwell 36. It lies just outside the eastern edge of the open cluster Melotte 111. Caldwell 38, or NGC 4565, whichever you prefer, can easily be found in any telescope of 80 mm and larger. But do not waste your edge-on galaxy nights with something small; look at this one with the biggest scope you can get hold of. Through an old 49 cm Newt. it looks just awesome, resembling a ghostly alien flying saucer coming into land; the central bulge and dust lane are easily visible, along with a 13th magnitude star just above (NE) of the bulge.

The superb edge-on galaxy NGC 4565 imaged by Gordon Rogers of Long Crendon, Aylesbury, UK with a 0.4 m Schmidt-Cassegrain at *f*/6.3 plus SBIG ST8E CCD and AO7 adaptive optics unit. http://www.gordonrogers.co.uk

Caldwell 39

NGC 2392
Eskimo Nebula in Gemini
Magnitude: 9.2
Size: 45″
Mag/Sec: 17
RA: 07 h 29 min 11 s
Dec: +20° 55′
Circumpolar: Above 70°N
Starhop: 7°S and 3½°W from Pollux (ß Gem)
Best Months: December to March
Distance: ~4,000 light-years
Physical Size: ~10 light-months
Age: ~1,000 years (inner) to ~5,000 years (outer)
Best Visual Aperture: 10 cm and larger with high powers
Best Visual Filter: OIII and UHC filters work well
CCD/DSLR Advice: Small, so 1″/pixel and finer works best
Celestial Neighbors: Open cluster NGC 2420 2½°ENE
Miscellaneous: A "must-see" planetary nebula
Chart: See Caldwell 25

Eastern Gemini is our next destination as we track down an object that really does resemble its popular name of the Eskimo. In any decent amateur image of NGC 2392, this planetary nebula actually looks like someone's face surrounded by the fur-lined hood of an anorak. So, in this case, the observer at the eyepiece may well resemble the object being looked at, especially as this object is definitely a winter feature in the northern hemisphere!

Sadly, the Eskimo face cannot be seen visually: the subtle patterns are just too faint. However, if you have a vivid imagination you might be able to convince yourself you are looking at an anorak hood with a nose in the center! If you are imaging it try taking a vanity mirror outside so you can compare your appearance with the image when it downloads. Like all the planetary nebulae we have encountered so far, the Eskimo is small but bright. Pollux is the nearest bright star, but third magnitude Wasat (δ Geminorum) is the closest obvious naked eye star to hop from. From Wasat you just go 2° east and 1° south. As always with planetary nebulae, ludicrously high magnifications with big telescopes will surrender the most detail to the naked eye, and there is plenty of detail to be seen. Even with small apertures (100 mm or less) the Eskimo reveals a bright center, an inner ring, and an outer halo. Look at it with a 40-cm aperture at 10 p.m. local time in February (when it almost reaches 60° altitude from the UK), and you may well be mesmerized by the subtle details. The central star (HD 59088) is magnitude 10.5 but has (allegedly) been recorded as bright as 9.7. The inner ring has a diameter of barely 15 arcsec, and the outer halo, stretching out to more than 20 arcsec from the central star, shows subtle texture with a greenish hue.

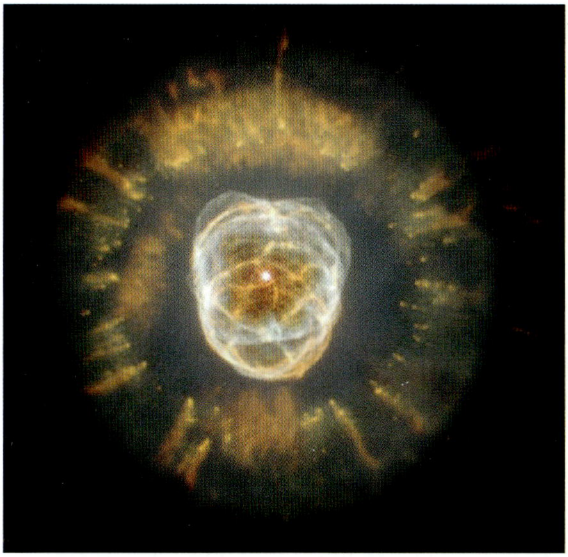

The Eskimo Nebula, NGC 2392, imaged in fine detail by the Hubble Space Telescope. The field is less than an arc-minute wide. Depending on your imagination it might look like an Eskimo, a Clown's face or even a Lion's face. Image: NASA/Andrew Fruchter/ ERO team.

Caldwell 40

NGC 3626
Galaxy in Leo
Magnitude: 11.0
Size: 3′ × 2′
Mag/Sec: 22
RA: 11 h 20 min 04 s
Dec: +18° 21′
Circumpolar: Above 72°N
Starhop: 4°N and 7°W from Denebola (ß Leo)
Best Months: February to May
Distance: 85 million light-years
Physical Size: ~75,000 light-years
Age: Billions of years
Best Visual Aperture: 25 cm and larger is recommended
Best Visual Filter: Deep sky filter may help
CCD/DSLR Advice: Try unfiltered imaging at long focal lengths
Celestial Neighbors: Many fainter galaxies nearby
Miscellaneous: Rather a drab object!

The Caldwell catalog contains some fantastic deep sky objects, but, sadly, Caldwell 40 is not one of them. Part of the problem lies in its considerable distance of 85 million light-years, but not all of it. Do not lose any sleep if you have not seen this object! This oval fuzz is simply not packed with intricate detail. Nor is it enormous and dazzlingly bright. Whether observed visually or with a CCD it is bland and as dull as dishwater.

Nevertheless, if you want to find it just search out that 10° wide triangle at the rear end of the Lion, composed of the stars Denebola, Zosma, and Chort, and, from Zosma move just more than 2° south and a degree and a half east. After a few seconds of viewing the galaxy you will be yawning continuously with tears rolling down your face, and the resulting dehydration might place you in a coma. Maybe Leo has more galaxies to offer, though? Well, obviously! What deep sky observer has not studied M65, M66, and NGC 3268 just more than 2° south and slightly east of Chort. Move a mere 8° to the west and three more tenth magnitude Messier galaxies are available, namely M95, M96, and M105. But maybe to keep on topic we should try to find something as close as possible to Caldwell 40 to retain our interest? Well, 1° to the west-southwest of Caldwell 40 are two similarly dull objects, which it is worth nudging the telescope towards. NGC 3607, at magnitude10.9, is actually brighter than the Caldwell galaxy. It also has a tiny 13th magnitude galaxy to its southwest. NGC 3608 is just to the north but is pretty drab. Frankly, if you want to view a great non-Messier galaxy in Leo, go right over to the western border with Cancer and bag NGC 2903. Visually or digitally, it is a stunner.

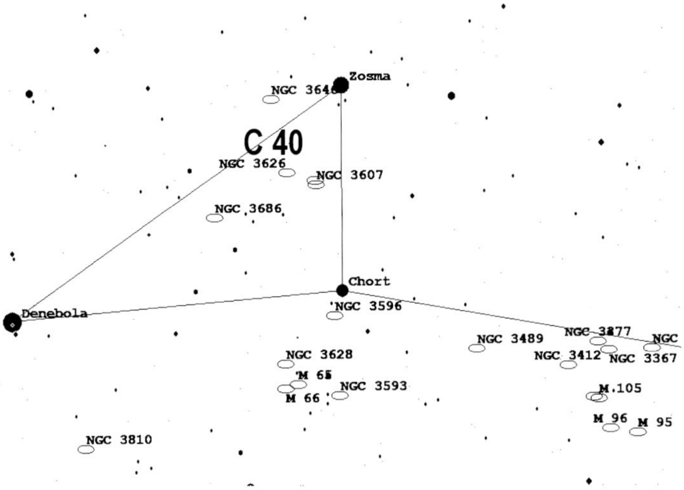

The rather drab galaxy NGC 3626 (C40) lies within the triangle of Denebola, Chort and Zosma at the eastern end of Leo. The far superior M65 and M66 are to the south, below Chort and M95, 96, and 105 are further west (*bottom right* in the chart).

NGC 3626 from the Digitized Sky Survey. The field is 3 arcmin wide.

Caldwell 41

Melotte 25
Hyades Cluster in Taurus
Magnitude: 0.5
Size: 5°
Mag/Sec: Not relevant
RA: 04 h 27 min
Dec: +15° 50′
Circumpolar: above 75°N
Starhop: Just find Aldebaran!
Best Months: November to February
Distance: 150 light-years
Physical Size: ~13 light-years
Age: ~600 million years
Best Visual Aperture: Binoculars work fine!
Best Visual Filter: No filter required
CCD/DSLR Advice: A DSLR and 200 mm lens works well!
Celestial Neighbors: Pleiades 12° to NW
Miscellaneous: Despite being C41, it has a declination lower than C43!

Every amateur astronomer has surely heard of the Pleiades and the Hyades, two northern hemisphere winter stellar groupings plainly visible to the naked eye. They are so obvious that one wonders whether they should be given extra publicity. However, if Messier felt the former object should be numbered "45" maybe that alone justifies a Caldwell designation for the Hyades. You will not need to hunt down Caldwell 41; just finding Aldebaran will do. To the eye the cluster is just a vague V shape, easier to see if Aldebaran is shielded from view. However, through a big pair of binoculars or a small telescope, the view is superb, even with the obviously red Aldebaran in the field. A really good instrument for this object is a Takahashi FS60c refractor, of 355 mm focal length, with a 32 mm eyepiece, giving 11×. The four and a half degree field is almost wide enough to fit the whole cluster in, and Aldebaran can be kept just outside the field if preferred. Excluding Aldebaran and a handful of third, fourth, and fifth magnitude stars, most of the stars in the cluster are sixth, seventh, and eight magnitude. Aldebaran lies only 65 light-years from Earth, whereas the cluster stars are 150 light-years distant. So, sadly, the vision that the "eye of the bull" is part of the Hyades is just a vision; Aldebaran is actually 20 light-years nearer to Earth than it is to the Hyades. From a planet going around Aldebaran the Hyades cluster would be almost twice as large as it appears to us. Looking in the opposite direction our Sun would be a feeble magnitude 6.3. It does not take a genius to work out that if Aldebaran shines at about magnitude 0.85 to us, but our Sun is almost 5.5 magnitudes fainter from that star, that the red giant is roughly 150 times more luminous. Aldebaran is thought to have a diameter almost 40 times that of our Sun. While in the region make sure you seek out the neglected magnitude 6 open cluster NGC 1647, just 3° to the northeast of Albebaran.

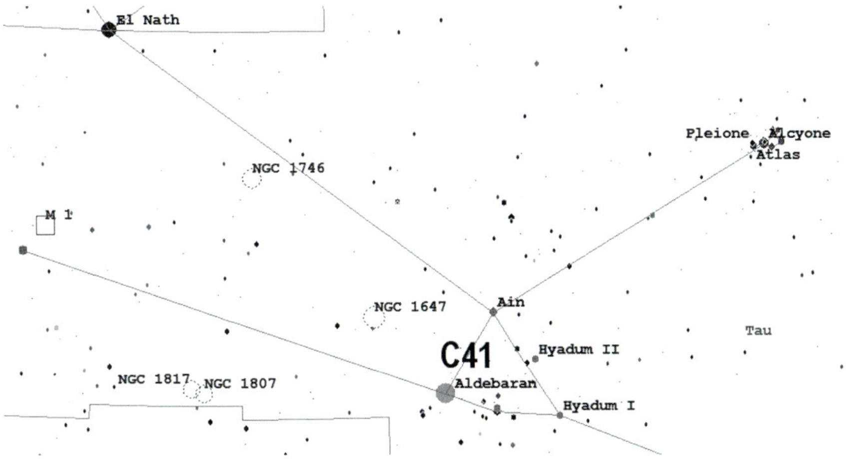

This finder chart for the Hyades is probably unnecessary as finding Aldebaran in Taurus is hardly a challenge. The V shaped cluster west of Aldebaran *is* the Hyades.

The Hyades star cluster photographed on 2000 January 25 by Till Credner (AlltheSky.com), from Lichtenhagen, Germany using a 135 mm *f*/2.8 lens, Kodak Royal Gold 400 film and a 30 min exposure. The field is 10° wide and the small open cluster NGC 1647 is visible in the top left.

Caldwell 42

NGC 7006
Globular Cluster in Delphinus
Magnitude: 10.5
Size: 3.5′
Mag/Sec: 22
RA: 21 h 01 min 30 s
Dec: +16° 11′
Circumpolar: above 74°N
Starhop: 7½°N and 17°E from Altair (α Aquilae)
Best Months: July to October
Distance: 135,000 light-years
Physical Size: ~140 light-years
Age: ~10 billion years
Best Visual Aperture: 20 cm and larger is recommended
Best Visual Filter: No filter advantage
CCD/DSLR Advice: Compact, but easy to image
Celestial Neighbors: Globulars C47 (10°SW) & M15 (7°SE)
Miscellaneous: Seriously neglected globular

Delphinus, the Dolphin, is a small constellation whose basic shape consists of a diamond of four stars, plus a tail of two, with none brighter than magnitude 3.7. But, despite this, the shape is easy to recognize, even from mediocre observing locations. The legendary comet and nova discoverer George Alcock found his first nova (HR Del) in Delphinus in 1967. It was an extraordinary object, and, even now, more than 40 years later, no other novae have *ever* been found in Delphinus. HR Del is, today, a 12th magnitude object still under study, situated in the far north of the constellation, just below the border with Vulpecula. Caldwell 42 is not far away, in the northeastern sector, close to the border with Pegasus. A magnitude 10.5 globular cluster at a distance of 135,000 light-years is never going to be a mind-blowing sight. Apart from the low surface brightness, more akin to that of a distant galaxy, the globular's individual stars are 15th magnitude and fainter, so only the largest telescopes in amateur hands will make NGC 7006 look like M13. It is easy to miss this globular in a small telescope at low powers. Locating fourth magnitude γ Delphini (the easternmost star in the diamond) and sliding 3½° east is the simplest way of finding NGC 7006. If you do not have "Go To" but do have setting circles the brighter Altair (α Aquliae) or Enif (ε Pegasi) can be used as starting points. If you find NGC 7006 a bit dull, and in a small telescope it is no more than a novelty to tick off your observing list, the sixth magnitude globular M15 is only 8° to the southeast. With a diameter of 12 arcmin and dozens of stars of 13th magnitude and brighter, it is a really globular, discovered by Jean-Dominique Maraldi in 1746, some 38 years before William Herschel found NGC 7006. But do not give up on Delphinus' globulars yet; there is a brighter example five objects later.

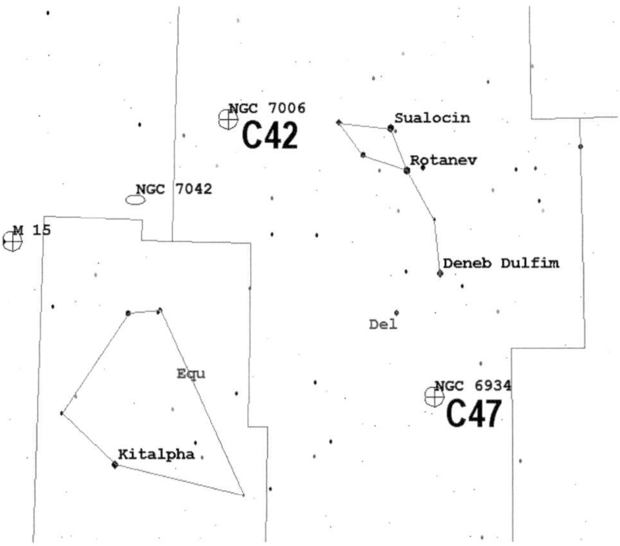

The tiny constellation of Delphinus to the northeast of Aquila contains two small globular clusters, NGC 7006 and NGC 6934, also known as Caldwell 42 and 47.

NGC 7006 imaged by the author on 2008 September 26 with a 0.35 m Celestron 14 at *f*/7.7 and an SBIG ST9XE CCD, 60 s exposure.

Caldwell 43

NGC 7814
Electric Arc Galaxy in Pegasus
Magnitude: 10.5
Size: 5.5′ × 2.5′
Mag/Sec: 23
RA: 00 h 03 min 12 s
Dec: +16° 09′
Circumpolar: above 74°N
Starhop: 1°N and 2.5°W from Algenib (γ Peg)
Best Months: August to December
Distance: 50 million light-years
Physical Size: ~80,000 light-years
Age: Billions of years
Best Visual Aperture: 20 cm and larger is recommended
Best Visual Filter: Deep sky filter may help
CCD/DSLR Advice: Well suited to deep monochrome imaging
Celestial Neighbors: Fainter galaxy NGC 14 is 80′ to ESE
Miscellaneous: Photographed by Keeler, from Lick, in 1899
Chart: See Caldwell 30

Caldwell's 43 and 44 are both tenth magnitude galaxies near the base of the famous "Square of Pegasus," as shown in the chart. So, for once, we can quickly slew between two consecutive Caldwell objects on the same night in the autumn sky. Caldwell 43, alias NGC 7814, is conveniently situated near the southeastern vertex of the square, namely, the star Algenib (γ Peg). Hopping a degree north and 2½° west from there gets you to the galaxy. If the galaxy is placed dead center in a 20 arcmin wide by 15 arcmin high CCD field a bright 7th magnitude star will just enter the northwest corner of the field. NGC 7814 is a healthy magnitude10.4, but when you first bag it the adjective "healthy" is the last one you would think of. This is a galaxy that can look stunning in a high-quality image with a large amateur telescope but can look drab through the eyepiece. In images the edge-on appearance and dark dust lane may remind you of a poor man's version of Caldwell 38/ NGC 4565. Take an image of that spectacular galaxy and lower the brightness, especially of the arms, make it smaller, but increase the central bulge, oh, and make the dust lane narrower and it almost turns into NGC 7814. Trouble is, you cannot do that visually. Look closely at a top-quality professional image and you will notice that the plane of the galaxy, indicated by the dust lane, is slightly warped and twisted. You also notice that many of the numerous background stars in the image are actually incredibly faint galaxies. Visually, NGC 7814 through a 35 cm Schmidt-Cassegrain at 200× looks very much like the Sombrero Galaxy, M104, seen through a 16 cm reflector at 200×. The name "Electric Arc" galaxy for NGC 7814 was coined by Lucien Rudaux (an artist and astronomer).

The fantastic edge-on spiral NGC 7814 imaged by Adam Block with the 20-in. (51 cm) *f*/8.4 RCOS Ritchey-Chrétien at the Kitt Peak visitors center. Paramount ME mount and SBIG ST10XME CCD and AO7 adaptive optics unit. One iteration of L-R deconvolution (sharpening) algorithm using CCDsharp was applied to the luminance image and Digital Development (DDP) via Maxim DL, and was also used to display the very dim and very bright details of the image simultaneously. LRGB 120:20:20:20 min. R,G, B binned 2 × 2. Image: Adam Block/NOAO/AURA/NSF.

Caldwell 44

NGC 7479
Galaxy in Pegasus
Magnitude: 10.8
Size: 4′ × 3′
Mag/Sec: 21
RA: 23 h 05 min 00 s
Dec: +12° 19′
Circumpolar: above 78°N
Starhop: 3°S from Markab (α Peg)
Best Months: August to November
Distance: 105 million light-years
Physical Size: ~120,000 light-years
Age: Billions of years
Best Visual Aperture: 20 cm and larger is recommended
Best Visual Filter: Deep sky filter may help
CCD/DSLR Advice: Well suited to deep monochrome imaging
Celestial Neighbors: Fainter galaxy NGC 7448 is 7°WNW
Miscellaneous: Twenty years since its last supernova
Chart: See Caldwell 30

Caldwell 44 is also conveniently situated near one of the Square of Pegasus' stars. This time it is Markab (α Peg) that we need to center, before moving almost 3° further south. Where Caldwell 43 was virtually edge-on, Caldwell 44 is face-on. It has a very distinctive shape, and as soon as you have seen an image of it you will never forget the galaxy's NGC designation of 7479. Maybe someone can explain this neural phenomenon! The shape is a barred spiral seen from above, always a fine sight, but, in this case, there are just two highly curved spiral arms emanating from the bar and arching themselves back in an anticlockwise direction around the galaxy. The result looks like a letter "S" lain on its side and reflected in a mirror. The westernmost arm has a larger radius and looks almost like a whip about to be cracked. A magnitude 13.7 foreground star in our Milky Way, looking tantalizingly like a supernova, sits in the gap between the western arm and the bar. The core of the galaxy is a bright knot in the center of that impressive bar. A real supernova has been discovered in NGC 7479. On July 27, 1990, Pennypacker and Perlmutter discovered a magnitude 16 Type Ic supernova quite close to the position of that imposter supernova. The best amateur images nicely reveal the golden hue of the central core and bar, gray and silvery blue aspects to the arms, and a smattering of pinkish HII regions as well. The vertical north-south aligned bar and the bright core are quite obvious under dark skies with a 16 cm instrument. In a 35 cm Schmidt-Cassegrain the S shape can be seen, although nowhere near as well as in a short CCD exposure with the same instrument.

NGC 7479 imaged by Gordon Rogers of Long Crendon, Aylesbury, UK with a 0.4 m Schmidt-Cassegrain at *f*/6.3 plus SBIG ST8E CCD and AO7 adaptive optics unit. http://www.gordonrogers.co.uk

Caldwell 45

NGC 5248
Galaxy in Boötes
Magnitude: 10.3
Size: 6′ × 4.5′
Mag/Sec: 22
RA: 13 h 37 min 30 s
Dec: +08° 53′
Circumpolar: above 82°N
Starhop: 10°S and 9°W from Arcturus (α Boötis)
Best Months: March to June
Distance: 75 million light-years
Physical Size: ~130,000 light-years
Age: Billions of years
Best Visual Aperture: 20 cm and larger is recommended
Best Visual Filter: Deep sky filter may help
CCD/DSLR Advice: Unfiltered and LRGB imaging work well
Celestial Neighbors: Globulars M53 & NGC 5053 are 10°WNW
Miscellaneous: Still awaiting a supernova discovery

Caldwell 45 lives in the constellation of Boötes, the herdsman, but only just. It is a scant third of a degree east of the constellation boundary with Virgo, the Virgin. In fact, it would be more appropriate if the galaxy were in Virgo, as it is a member of the giant Virgo Cluster. Draw a line between two of the brightest stars in the sky, namely Arcturus and Spica, and NGC 5248 lies to the northwest of the midpoint of that line. If you draw a giant triangle between Mufrid (η Boötis), Vindemiatrix (ε Virginis), and NGC 5248 the galaxy is almost equidistant between those two stars and sits at the right angle of the triangle. Unfortunately, there are no other distinctive star patterns to help you home in on the 10th magnitude galaxy as it lie in a particularly barren region of the sky with only 5th or 6th magnitude stars (at best) between it and those aforementioned bright stars. Through a 16 cm reflector, the elliptical shape of the galaxy is obvious, with a bright core surrounded on either side by what look vaguely like parentheses. A single star of magnitude 14 sits just outside the fuzzy southern border of the galaxy in larger apertures.

CCD images of the almost face-on NGC 5248 are, to some at least, a bit reminiscent of images of that fine galaxy M100, except this Caldwell galaxy looks like it has been stretched or squashed, and the spiral arms themselves are trying to go the way of the reflected "S" shape of NGC 7479. In color images two main steely blue arms dotted with pink HII regions make the galaxy a prime target for those with large apertures and LRGB imaging experience.

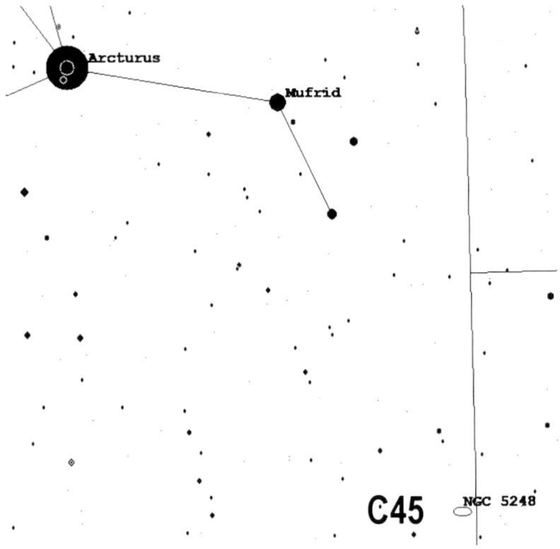

The chart shows the location of Caldwell 45, close to the Boötes/Virgo border, in relation to the brilliant star Arcturus.

NGC 5248 from the Digitized Sky Survey. The field is 10 arcmin wide.

Caldwell 46

NGC 2261
Hubble's Variable Nebula in Monoceros
Magnitude: ~10
Size: 3.5′ × 1.5′
Mag/Sec: 20
RA: 06 h 39 min 12 s
Dec: +08° 44′
Circumpolar: above 82°N
Starhop: 1½°N and 11°E from Betelgeuse (α Orionis).
Best Months: November to February
Distance: 2,500 light-years
Physical Size: ~30 light-months
Age: 300,000 years?
Best Visual Aperture: 15 cm and larger is recommended
Best Visual Filter: Deep sky filter
CCD/DSLR Advice: Animations worthwhile (see text)
Celestial Neighbors: Rosette (C49/50) lies 4° to SSW
Miscellaneous: Great CCD animation target

Hubble's Variable Nebula in Monoceros is one of those objects that many amateurs recall seeing an image of, but they are never quite sure where to find it or indeed what it is! Perhaps its proximity to the Rosette Nebula (see Caldwell's 49 and 50) and the deep sky gems in Orion tend to steer amateurs away from this remarkable object. Also, perhaps there is a subconscious view that it is probably far more difficult to spot than it actually is. In fact, in any modest telescope, it is an easy object. Everyone can find the giant red star Betelgeuse in the northeastern corner of Orion; from there you just move a degree and a half north and 11° east.

When you first spot the Nebula it looks like a tenth magnitude comet, as it has a distinctive fan-shaped tail. At the southern apex of the nebula is a bright point that looks like a star. This object is the variable star R Monocerotis, although it is actually a star surrounded by several Earth-masses of dust. Infrared images can peer through the dust and see the actual star, but amateurs are looking at a glowing dust cloud that varies between magnitude 11.0 and 13.8. Both the starlike object at the tip and the nebula itself vary in brightness. However, the variations are *not* coincidental. In fact, it would be rather worrying if they were, as the nebula is at least 30 light-months in length if it is face-on to us. Visual observers can sometimes detect changes in the nebula's appearance in timescales of weeks or months. However, CCD imagers can really go to town on this object and create an animation of the nebula's light patterns changing, or flickering like a candle flame. What appears to be happening is that dust clouds close to the hidden R Mon. star occasionally break and allow more light from the hidden 5,500 solar luminosity star to escape and shine toward the nebula, illuminating the regions we see many weeks, months, or years later. Definitely an object worth monitoring, as there are only four other variable nebulae: Caldwell 68, Hinds', McNeil's, and Gyulbudaghian's.

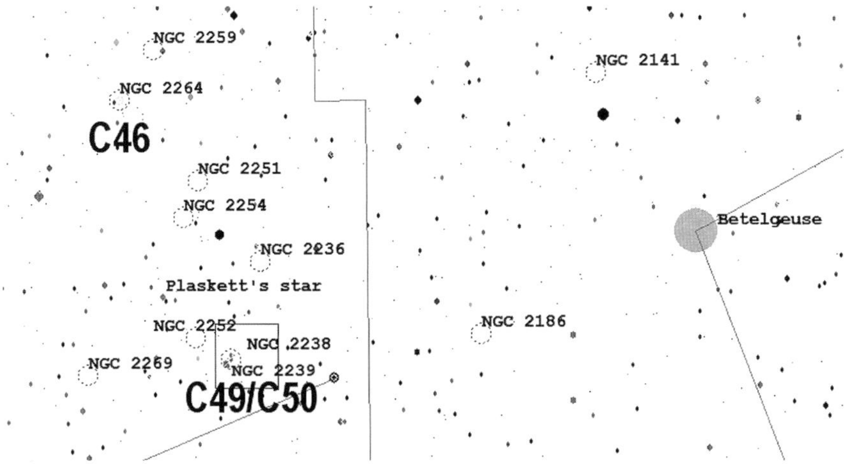

Both Caldwell 46 and Caldwell 49/50 (the Rosette) live in western Monoceros not far from the well known naked eye star Betelgeuse in Orion's northeast corner.

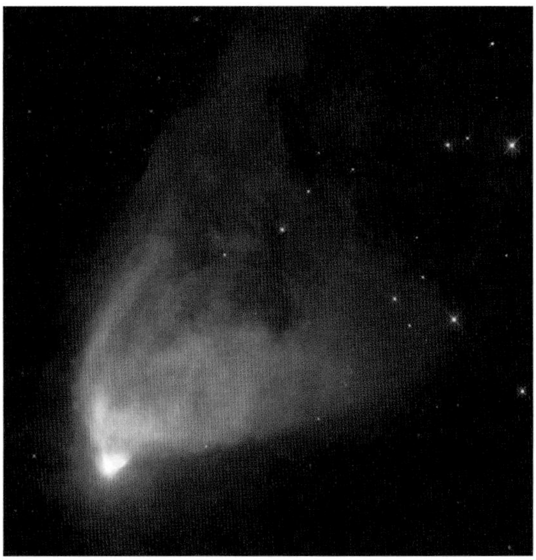

Hubble's variable nebula NGC 2261, imaged in high resolution by the Hubble Space Telescope. Image: Hubble Heritage Team – AURA/STScI/NASA.

Caldwell 47

NGC 6934
Globular Cluster in Delphinus
Magnitude: 8.8
Size: 7′
Mag/Sec: 21
RA: 20 h 34 min 12 s
Dec: +07° 24′
Circumpolar: above 83°N
Starhop: 1½°S and 10¾°E from Altair (α Aquilae)
Best Months: July to October
Distance: 60,000 light-years
Physical Size: ~120 light-years
Age: ~10 billion years
Best Visual Aperture: Easy with 10-cm apertures and larger
Best Visual Filter: No filter advantage
CCD/DSLR Advice: Easy to image with any detector
Celestial Neighbors: Globulars C42, M15 & M2 within 15°
Miscellaneous: Almost as neglected as NGC 7006!
Chart: See Caldwell 42

NGC 6934 is the second globular cluster in Delphinus. We encountered the dimmer, northern example (NGC 7006) five objects earlier, but because the Caldwell catalog is listed in descending declination order, dropping 9° has let four other non-Delphinus objects barge into the queue.

There are various ways you could get to NGC 6934. It is almost 11° east and a degree and a half south of Altair, but alternatively you can just follow the curve of the Dolphin south, fall off the back of magnitude 4.0 Deneb Dulfim (ε Delphini), and travel 4° further.

Caldwell 47 is a much more pleasing object than Caldwell 42, simply because the latter is more than twice as far away. An object a full 7 arcmin in diameter is much easier to find in a small telescope at low powers, so hunting Caldwell 47 down is not too tricky. Through a telescope the first distinguishing feature of this globular is the presence of the magnitude 9.2 star GSC 5222249 only 2 arcmin west of the cluster's core. This star is, of course, within our own Milky Way Galaxy and is thought to lie at a distance of 160 light-years, although this distance is very uncertain. Nevertheless, the globular is hundreds of times more distant and, at roughly 120 light-years across is almost as large as the foreground star's distance from us. The brightest stars in the globular itself are, at best, 13th magnitude, and most are 14th to 15th. However, this does mean that with a half meter-sized Dobsonian NGC 6934 will come alive. The NASA Astronomy Picture of the Day (APOD) for June 30, 1999, was a remarkable image of NGC 6934 taken with the 8.1 meter Gemini North telescope, an infrared detector, and the University of Hawaii's Hokupa'a Adaptive Optics System. The picture had an astonishing resolution of 0.09 arcsec, making the densely packed globular look a rather more open place!

The globular cluster NGC 6934 imaged by the author on 2008 September 26 with a 0.35 m Celestron 14 at *f*/7.7 and an SBIG ST9XE CCD, 60-s exposure.

Caldwell 48

NGC 2775
Galaxy in Cancer
Magnitude: 10.0
Size: 4.6′ × 3.8′
Mag/Sec: 22
RA: 09 h 10 min 18 s
Dec: +07° 02′
Circumpolar: above 83°N
Starhop: 5°S and 14½°W from Regulus (α Leo)
Best Months: January to April
Distance: 55 million light-years
Physical Size: ~75,000 light-years
Age: Billions of years
Best Visual Aperture: 15 cm and larger
Best Visual Filter: Deep sky filter may help
CCD/DSLR Advice: Complicated! (see text below)
Celestial Neighbors: Cluster M 67 lies 5° to NW
Miscellaneous: Incredibly delicate spiral arms

Draw a line between the bright star's Regulus (in Leo) and Procyon (in Canis Minor) and Caldwell 48 lies just south of that line and slightly nearer to Regulus. However, it is quite a haul from either star, and a better philosophy would be to locate the two third magnitude stars at the very top of Hydra (2° apart, right near the border with Cancer) and then hop from the easternmost one (ζ Hydrus) almost 4° east and a little more than 1° north, crossing the Hydra/Cancer border in the process.

This is yet another Caldwell galaxy that looks far better in amateur images than it does through the eyepiece of even a large backyard telescope. At first sight it looks like it is face-on, but it is actually tilted almost 40° away and so appears elliptical in shape. The galaxy has an unusual but fascinating appearance. In some respects it looks a bit like a simpler version of the Sunflower Galaxy, M63, in that it has a large flat and bright central disc that abruptly changes into some highly detailed but fragile spiral arms. However, in the case of Caldwell 48 the word "fragile" could be replaced by "almost invisible." The spiral arms are incredibly delicate and tightly wrapped; indeed, they look more like concentric rings than arms, even in professional images. In the very best amateur CCD images they can just be resolved as faint gray spirals, but only just: adaptive optics may help here. Even with a large amateur telescope you will see just a bright core visually and a very faint elliptical halo. There is little color in the galaxy, just a central creamy/brown, fading to gray in the arms, so monochrome images do not lose you much color detail. In half-degree tall amateur images you will capture the fainter 14th magnitude galaxies NGC 2777 and NGC 2773 and maybe 16th magnitude PGC 0213576 as well. One supernova has been found in Caldwell 48, a 13th magnitude Type Ia discovered in late September 1993 by Treffers, Fillipenko, et al.

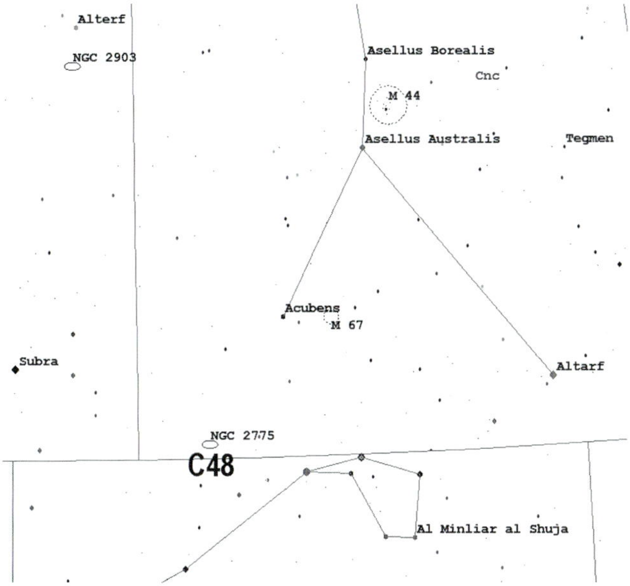

NGC 2775 (C48) lies in southeastern Cancer close to the borders with Hydra and Leo.

This excellent image was taken by Philipp Keller, Christian Fuchs and the T1T team using the Trebur 1.2 m Cassegrain at Trebur, Germany. The telescope was used with a 0.5× focal reducer giving 5.1 m focal length (f/4.25) and an SBIG ST10E CCD. LRGB: 40:10:10:10 min. Luminance binned 2 × 2. R,G, B, binned 3 × 3. All exposures were stacks of individual 2 min unguided exposures. The image was stretched hard to make the outer dust band visible. Note the group of tiny galaxies just north of NGC 2775. Image by kind permission of Philipp Keller.

Caldwell 49 and Caldwell 50

NGC 2237/8 and 2246
Rosette Nebula in Monoceros and Open Cluster NGC 2244
Magnitude: ~5 for each
Size: ~80′ (nebula) and 30′ (cluster)
Mag/Sec: 23–24
RA: 06 h 32 min 00 s
Dec: +05° 00′
Circumpolar: above 85°N
Starhop: 2½°S and 9°E from Betelgeuse (α Orionis)
Best Months: November to February
Distance: 5,000 light-years
Physical Size: ~120 light-years
Age: Cluster stars ~1 million years
Best Visual Aperture: Nebula is v. elusive (see text below)
Best Visual Filter: Greatly enhanced by OIII and UHC
CCD/DSLR Advice: Wide field H-alpha shots work best
Celestial Neighbors: C46 and numerous open clusters
Miscellaneous: Big and beautiful CCD target
Chart: See Caldwell 46

Ten degrees east and some 2½° south of the giant presupernova candidate star Betelgeuse in Orion is one of the largest and most popular CCD and DSLR targets in the night sky: the Rosette Nebula in Monoceros. The Rosette (Caldwell 49) is more than a degree in diameter and consists of the NGC objects 2237, 2238, and 2246. The big problem is that to most visual observers with light-polluted skies it is invisible, not only to the naked eye but with any form of optical aid, unless OIII or UHC filters are used. The Orion Nebula (just a short distance away) blows it away a hundredfold. Some may disagree with this assessment, but prior to E.E. Barnard's photography of the feature in the 1890s there are just a few isolated glimpses of the brightest parts of the Rosette by the greatest visual observers of the late eighteenth and nineteenth centuries, and skies were *much* darker in those days.

However, there is another aspect to the Rosette Nebula that can easily be seen visually; the glittering object at its heart, the open cluster NGC 2244. Patrick chose to catalog this cluster independently, as Caldwell 50. NGC 2244 is a fine sight in any telescope. In long exposure images it lies almost in the middle of the hole in the Rosette shape and consists of a dozen stars between sixth and ninth magnitude, together with dozens more down to twelfth magnitude. Technically, 12 Monocerotis, the brightest star, is not actually a cluster member, but it looks like it is. The entire cluster spans half a degree, or the same diameter as the Moon. NGC 2244 can be glimpsed as a fuzz with the naked eye, and the inexperienced observer may think he or she is seeing the Rosette. However, the nebula itself is almost three times the diameter of the cluster and definitely more impressive in a digital image than visually.

A color stack of forty-eight 300 second images of the Rosette nebula and cluster by Ian Sharp with a Takahashi Epsilon 130 mm Astrograph (*f*/3.6) and Hutech-modified Canon 350D on an Astrophysics AP1200 GTO mount. Images taken in 2008 February.

Caldwell 51

IC 1613
Scarecrow Galaxy in Cetus
Magnitude: 9.2
Size: 19′ × 17′
Mag/Sec: 24
RA: 01 h 04 min 50 s
Dec: +02° 07′
Circumpolar: above 88°N
Starhop: 13°S and 13°E from Algenib (γ Peg)
Best Months: September to December
Distance: 2.3 million light-years
Physical Size: 13,000 light-years
Age: Billions of years
Best Visual Aperture: 20 cm and larger is advised
Best Visual Filter: Deep sky filter may help
CCD/DSLR Advice: Deep unfiltered images work best
Celestial Neighbors: Many fainter galaxies to the east
Miscellaneous: 1906 photographic discovery by Max Wolf

Caldwell 51 is the last Caldwell object to reside in the northern celestial hemisphere. The remaining 58 deep sky objects therefore live in the south. From northern latitudes, the window of opportunity for viewing objects on the celestial equator at a decent altitude of 30° is less than 6 h, even in the long winter nights. Summer objects suffer badly as twilight virtually writes the entire month of June off and hampers most deep sky plans from mid-May to the end of July. Those observers who live further south will have a much happier time reading the rest of this book.

When you first see a picture of the irregular barred dwarf galaxy IC 1613 and read that the heap really is a galaxy you cannot help thinking that this is some sort of a sick joke. It just looks like a pile of rubbish. Where is the central core and where are the spiral arms? When you realize it is only 2.3 million light-years away you wonder whether it can seriously be called a galaxy at all. You even start to wonder why it is in the Caldwell catalog, although it certainly has a novelty value to it. This is not a galaxy that you are going to rush outside to image; it has no beauty at all! In terms of distance, size, and lack of beauty, it has a lot in common with Caldwell's 17 and 18 (NGC's 147 and 185), but it is even uglier and harder to see. The nearest bright star is Algenib in the southeast corner of the Square of Pegasus, but if you can find fourth magnitude ε Piscium and hop 5¾° south and ½° east you might be unlucky enough to spot it in the eyepiece. Our advice: Find something nicer to look at.

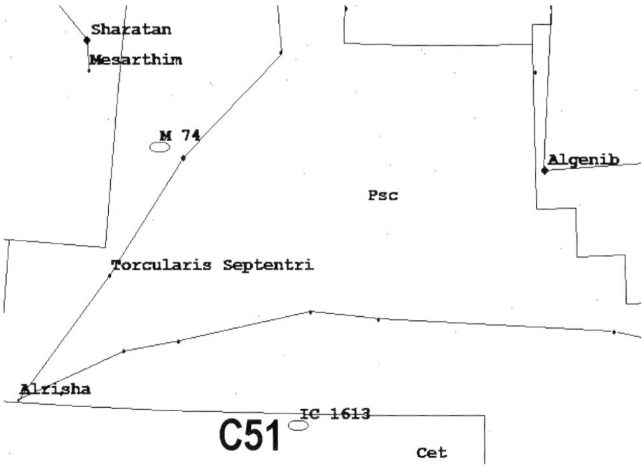

Caldwell 51 lies in a fairly barren region in northern Cetus, very close to the border with Pisces. The southeastern "Square of Pegasus" star Algenib is not far away.

IC 1613 from the Digitized Sky Survey. The field is 15 arcmin wide.

The Caldwell Objects

Caldwell 52

NGC 4697
Galaxy in Virgo
Magnitude: 9.2
Size: 7′ × 5.5′
Mag/Sec: 22
RA: 12 h 48 min 36 s
Dec: –05° 48′
Circumpolar: below 85°S
Starhop: 5½°N and 9°W of Spica (α Virginis)
Best Months: March to May
Distance: 75 million light-years
Physical Size: ~150,000 light-years
Age: Billions of years
Best Visual Aperture: 15 cm and larger
Best Visual Filter: Deep sky filter may help
CCD/DSLR Advice: Deep unfiltered images work best
Celestial Neighbors: M 104 lies 5° to the SSW
Miscellaneous: Massive elliptical galaxy

Draw a line from Spica, heading northwest to the famous double star Porrima (γ Virginis), and half way, if you branch off to the west, you may encounter NGC 4697, the first Caldwell object in the southern celestial sphere. Roughly at that half way point you may spot, with the naked eye, the magnitude 4.4 star θ Virginis. From there NGC 4697 is slightly more than 5° to the west.

Being Virgo there are lots more galaxies in the general region, but Caldwell 52 is the brightest one in the immediate vicinity. You need to go 6° south-southwest to M104, or12° northwest to M61 to encounter any serious galactic rivals. One minor pointer to the field of Caldwell 52/NGC 4697 is a magnitude 6.3 star (GSC 49551287), slightly more than half a degree to the galaxy's south-southwest.

NGC 4697 is an elliptical (or lenticular, opinions vary) galaxy of significant size. Assuming its distance of 75 million light-years is correct, it has a diameter of 150,000 light-years, slightly bigger than the Andromeda Galaxy. Admittedly, next to Perseus A (200,000 light-years in diameter, see Caldwell 24), Caldwell 52 looks rather tame, but it is 50% bigger than our own Milky Way Galaxy. But, there is more news in terms of its mass. In recent years professional astronomers have concluded NGC 4697 has a 50–100 million solar mass black hole at its core.

Through any telescope of 20-cm aperture or larger the galaxy resembles a fuzzy lenticular walnut, but with really dark skies and imaging skills a certain amount of luminous beauty can be revealed, as shown in the excellent accompanying image by Daniel Verschatse.

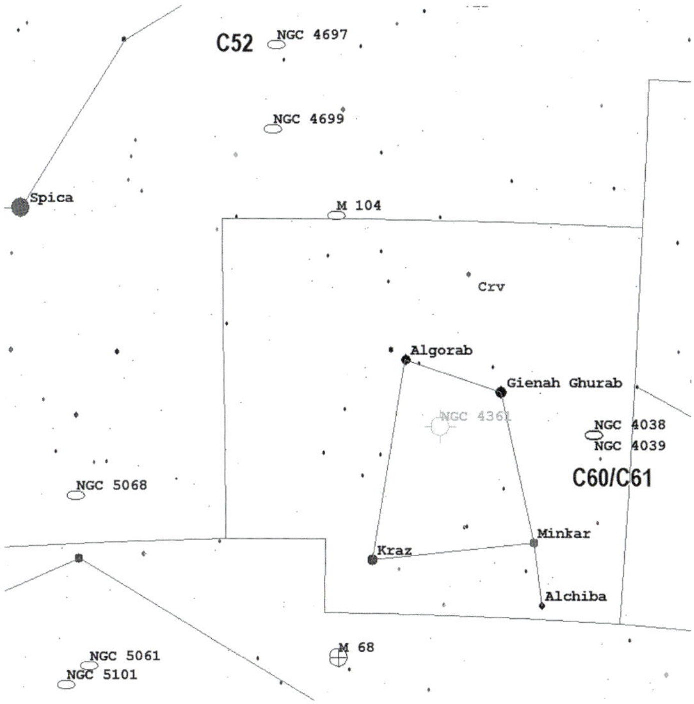

The chart shows NGC 4697 (C52) in western Virgo as well as NGC 4038/4039 (C60/61), further south in Corvus.

A splendid image of NGC 4697 taken by Daniel Verschatse at his Observatorio Antilhue, San Esteban, Chile. RCOS 14.5-in. (37 cm) Ritchey-Chrétien @ *f*/9 (prime focus). Astro-Physics AP1200-GTO mount. SBIG ST-10XE self-guided CCD. LRGB 60:20:20:20 min @ –15°C. 2005 April 10 @ 04:00 UTC. © Daniel Verschatse - Observatorio Antilhue – Chile.

Caldwell 53

NGC 3115
Spindle Galaxy in Sextans
Magnitude: 8.9
Size: 7.0′ × 3.5′
Mag/Sec: 21
RA: 10 h 05 min 12 s
Dec: −07° 43′
Circumpolar: below 83°S
Starhop: 1°N and 9½°E from Alphard
Best Months: January to April
Distance: 20 million light-years
Physical Size: ~40,000 light-years
Age: Billions of years
Best Visual Aperture: 15 cm and larger
Best Visual Filter: Deep sky filter may help
CCD/DSLR Advice: Deep unfiltered CCD images work best
Celestial Neighbors: Many fainter galaxies in the area
Miscellaneous: 75 years since its last supernova

Almost 20° due south of the bright star Regulus (the dot in the backwards question mark of Leo's head) you will find the Spindle Galaxy, in the small constellation of Sextans, the Sextant. Sextans is one of the drabbest constellations known. Although its official constellation boundary is square, the sextant shape itself is a triangle comprising one-fourth magnitude and two-fifth magnitude stars. The Spindle lies to the east (and slightly north) of the southernmost triangle star γ Sextantis. However, at a pitiful magnitude 5.1, most observers may prefer a brighter marker to the region, and locating Alphard (α Hydrus) to the west makes life easier. From there Caldwell 53 is 9½° east and 1° north, and you pass by γ Sextantis en route. The Spindle looks surprisingly bright for a non-Messier galaxy. The nearest galaxies that compete with it in that part of the sky are NGC 3521, almost 20° to the northeast and M95 plus M96 20° to the north-northeast.

The Spindle is another lenticular-shaped galaxy, and we see it edge-on so it looks a bit like a defocused NGC 891 (Caldwell 23), surrounded by two levels of fainter outer halos in images. In apertures below 15 cm, the Spindle may be unimpressive in hazy skies, but above 20 cm in crystal clear skies it is well worth seeking out. The galaxy core is quite intense, with the whole central bulge taking on a ghostly elliptical eye shape in large apertures. In April 1935 a professional astronomer named Samaha discovered a supernova roughly 1 arcmin north-northwest of the core of the Spindle, and it was designated 1935B, although there seem to be precious few details about its magnitude or type.

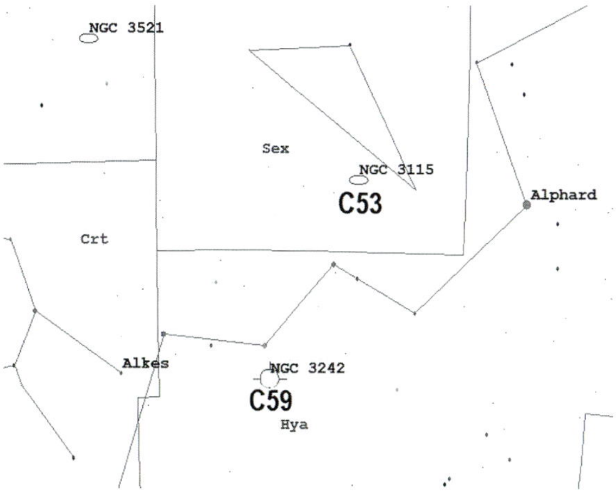

The chart shows the location of NGC 3115 (C53) in Sextans as well as NGC 3242 (C59) in Hydra.

A splendid image of NGC 3115 taken by Daniel Verschatse at his Observatorio Antilhue, San Esteban, Chile. RCOS 14.5-in. (37 cm) Ritchey-Chrétien @ *f*/9 (prime focus). Astro-Physics AP1200-GTO mount. SBIG ST-10XE self-guided CCD. LRGB 60:20:20:20 min @ –15°C. 2005 April 10 @ 00:30 UTC. © Daniel Verschatse - Observatorio Antilhue – Chile.

Caldwell 54

NGC 2506
Open Cluster in Monoceros
Magnitude: 7.6
Size: 10′
Mag/Sec: Not relevant
RA: 08 h 00 min
Dec: –10° 46′
Circumpolar: below 80°S
Starhop: 6°N and 18°E from Sirius (α Canis Majoris)
Best Months: December to February
Distance: 10,000 light-years
Physical Size: ~30 light-years
Age: ~3.5 billion years
Best Visual Aperture: 10 cm and larger
Best Visual Filter: No filter required
CCD/DSLR Advice: Easy to image with any telescope
Celestial Neighbors: Open cluster M48 lies 5° to the NE
Miscellaneous: Just one of many Monoceros clusters

You may have thought that, after the Rosette Nebula's fine open cluster (Caldwell 50), we would not be seeing any more clusters in the constellation of Monoceros. Think again! We are much further south and west this time, in the extreme southwest corner of Monoceros, in fact, but not quite far enough south to encroach into Puppis, although it is right near the border. The whole of southern Monoceros, northeastern Canis Major, and, especially, Puppis is stuffed to the brim with open clusters; many are as bright, or brighter than, NGC 2506. The best known open clusters in this part of the sky are M48 and M50, which are almost two magnitudes brighter and significantly bigger. If you imagine a large right angled triangle with Sirius and Procyon at the two acute angle vertices, NGC 2506 sits at the right angle. Finding the cluster in binoculars can be achieved by finding the surprisingly dim (magnitude 3.9) α Monocerotis and hopping one and a quarter degrees south and four and three-quarter degrees east. If we assume Caldwell 54 has a diameter of 10 arcmin (there seem to be a variety of opinions), then only the central 5 arcmin seems to have any real density to it. In a 30 cm aperture this "core" has a sort of upside-down "T" shape (with south at the top) and a triangle of 11th and 12th magnitude stars at the eastern edge. The base of the "T" also features two stars of about magnitude 11. Apart from these brightest stars several dozen 13th to 14th magnitude stars make up the rest of the cluster. With a telescope such as a large Dobsonian these fainter stars can be easily resolved and held steady as points, rather than a fuzz, greatly enhancing the view of Caldwell 54. Users of such apertures might like to swing them a degree and a half southeast from the cluster to see if they can detect the 12th magnitude galaxy NGC 2525.

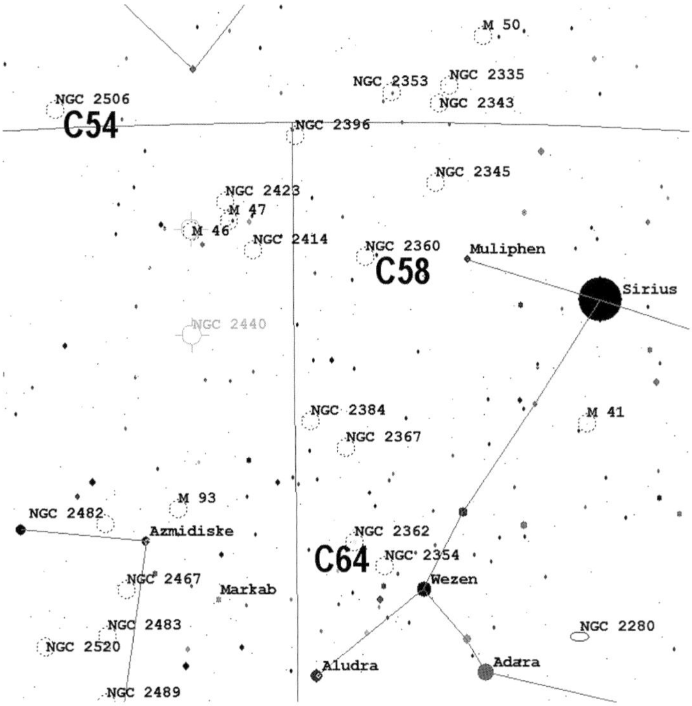

Caldwell's 54, 58, and 64 (NGC's 2506, 2360, and 2362) reside in southern Monoceros and Canis Major.

NGC 2506 from the Digitized Sky Survey. The field is 15 arcmin wide.

Caldwell 55

NGC 7009
Saturn Nebula in Aquarius
Magnitude: 8.0
Size: 45″ × 25″
Mag/Sec: 15
RA: 21 h 04 min 11 s
Dec: −11° 22′
Circumpolar: below 79°S
Starhop: 6°S and 6¾°W from Sadalsuud (ß Aqr)
Best Months: July to September
Distance: 1,400 light-years
Physical Size: ~15 light-weeks
Age: ~6,000 years
Best Visual Aperture: 15 cm and larger with high powers
Best Visual Filter: UHC or OIII filter works best
CCD/DSLR Advice: Good tracking at <1″/pixel desirable
Celestial Neighbors: Clusters M72 (glob) & M73 (open) 2°SW
Miscellaneous: Looks like a star at very low powers

The constellation of Aquarius, the Water Bearer, transits at local midnight in the month of August, invariably the warmest month of the year in the north and just free of the permanent twilight nights of June and July. Thus, despite the fact that Caldwell 55 can never even reach 30° altitude from, say, Suffolk in the UK, it can, at least, be viewed in a dark sky and in pleasant temperatures. It also has two very nearby and much more obvious Messier neighbors in the forms of the open cluster M73 and the globular cluster M72, so a trip to the far south of Aquarius can be very rewarding.

To find the region in question just start at the bright star Altair in Aquila and, moving 25° to the southeast, locate the magnitude 2.9 star Sadalsuud in Aquarius (β Aqr). Our target, the Saturn Nebula, alias NGC 7009 or Caldwell 55, is 6° further south and just under 7° west of Sadalsud. If you overshoot the field, you will probably run into M72 or M73, and so these are excellent markers.

Like all the planetary nebulae we have encountered the Saturn Nebula is small and has a high surface brightness. But, at low powers, it is easy to sweep over it, as it looks like a star. So sweeping the field at 50× or higher will reveal its nature. Do not expect the nebula to look like Saturn visually; the appearance of Saturn with its rings edge-on only becomes apparent in images. Visually, especially at low powers, it has a definite blue-green cast and looks like the ghost of a blue planet (Earth maybe?) or even an elliptical eye. The central star, at magnitude 12, will not dazzle the observer, either. Two faint blobs at the ring ansae are all you can really hope to see of the rings. Depending on your experience and the quality of your skies, anything from a 25 to a 50-cm instrument will be needed to see these blobs. The Saturn Nebula was observed by this author on September 26, 2008, with a Celestron 14 and was imaged shortly afterwards. Exposures of a second or so with the image contrast reduced to a minimum best represented what could be seen with the same instrument at 500×.

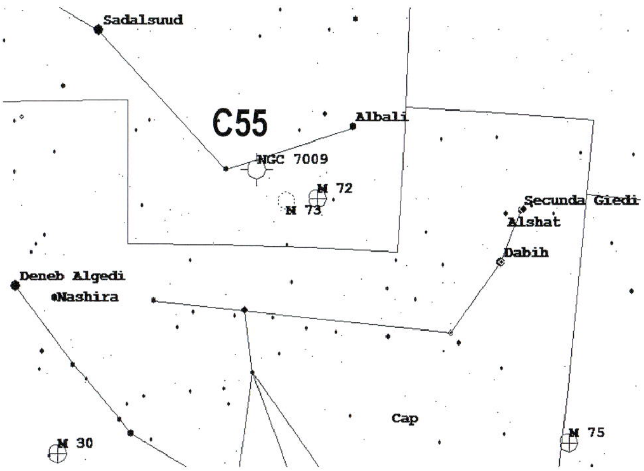

The chart shows the location of Caldwell 55 in southern Aquarius just north of the Capricornus border.

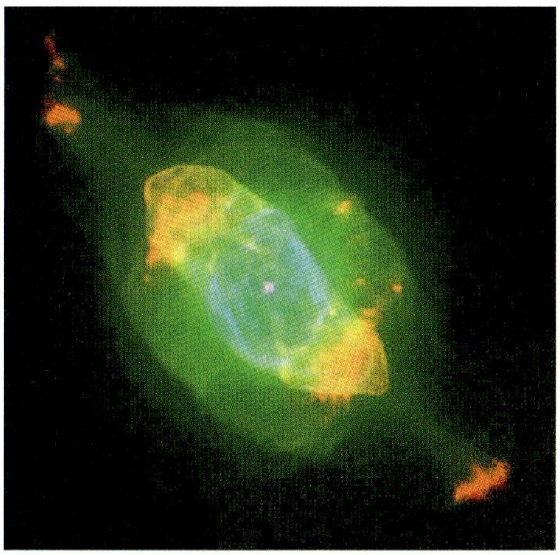

A spectacular high power view of the Saturn Nebula taken by the Hubble Space Telescope. Credit: Bruce Balick (University of Washington), Jason Alexander (University of Washington), Arsen Hajian (U.S. Naval Observatory), Yervant Terzian (Cornell University), Mario Perinotto (University of Florence, Italy), Patrizio Patriarchi (Arcetri Observatory, Italy), NASA/ESA.

Four co-added two second exposures by the author, enlarged considerably, show roughly how the Saturn Nebula looks visually through a large telescope. 0.35 m Celestron 14 at f.7.7 plus SBIG ST9XE.

Caldwell 56

NGC 246
Planetary Nebula in Cetus
Magnitude: 10.9
Size: 4.5′ × 4.0′
Mag/Sec: 22
RA: 00 h 47 min 03 s
Dec: −11° 52′
Circumpolar: below 79°S
Starhop: 6°N and 1°E from Diphda (ß Ceti)
Best Months: September to November
Distance: 1,600 light-years
Physical Size: ~2 light-years
Age: Thousands of years
Best Visual Aperture: 20 cm and larger
Best Visual Filter: OIII filter works well
CCD/DSLR Advice: Deep unfiltered exposures work best
Celestial Neighbors: Galaxy C62 (NGC 247) lies 9° further S
Miscellaneous: Surprisingly large and deceptively faint!

The vital steppingstone for finding Caldwell 56 is locating the bright (magnitude 2.0) star Diphda (β Ceti), which sits at 0 h 43.5 min R.A. and −18.0 Dec. Diphda is at the southerly apex of the big triangle formed by itself and two fainter stars of magnitude 3.5, namely Deneb Algenubi and Deneb Kaitos Shemali. The planetary nebula is slightly northeast of this triangle's center, 6° north and a degree east of Diphda. Two truly magnificent galaxies lie to the south of Diphda, namely NGC's 247 and 253. We will meet them later as Caldwell's 62 and 65.

There seems to be quite a variation in the quoted magnitude of this nebula. Many sources list it at photographic magnitude 8.0, which would make it a very healthy deep sky object. However, its visual magnitude is 10.9, some 15 times fainter! The problem seems to lie with the fact that the nebula has a central star and an off-center star in the same field, only an arcminute to the southwest. Two arcminutes to the northwest of the central star, near the nebula's edge, lives another conspicuous star, and the whole region is bathed in nebulosity. The central star seems to vary from 12th to 11th magnitude (and maybe much brighter). The off-center star is of a similar brightness (magnitude 11.7), and the star on the northwest edge is the brightest, at roughly magnitude 11.2. Added together the light from these three stars amounts to tenth magnitude, an unwelcome complication when estimating a nebula's brightness. At 100× in a 25 cm Newtonian, even from northern latitudes, the nebula resembles a crescent or half moon shape with the eastern side missing, and those three stars are quite obvious. Stephen O'Meara calls the nebula the Pac-Man Nebula because of its vague resemblance to the Japanese arcade game character that resembles a chomping head. Others have called it the Skull Nebula, for obvious reasons. As the Pac-man moniker has also been used for NGC 281 in Cassiopeia, calling it "the Skull" is less confusing.

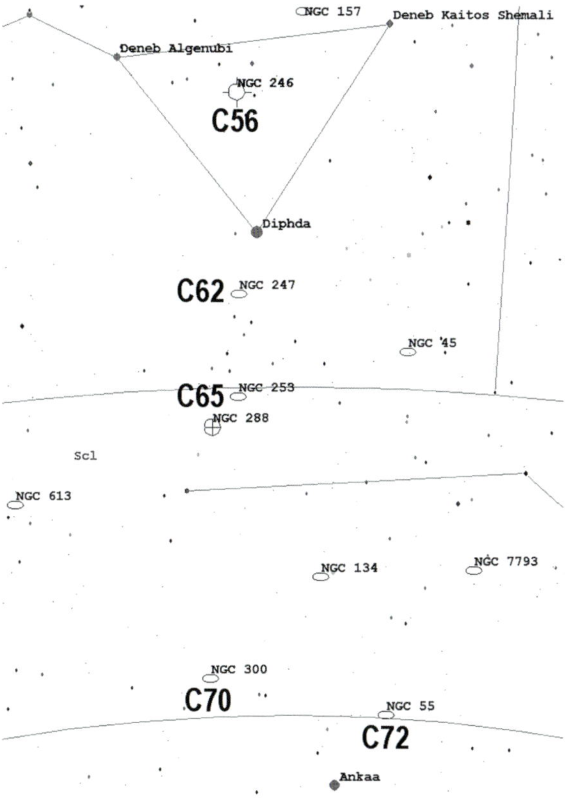

The chart shows the southern constellations of Cetus and Sculptor which contain the planetary nebula NGC 246 (C56) the four large galaxies NGC 247 (C62), NGC 253 (C65), NGC 300 (C70), and NGC 55 (C72).

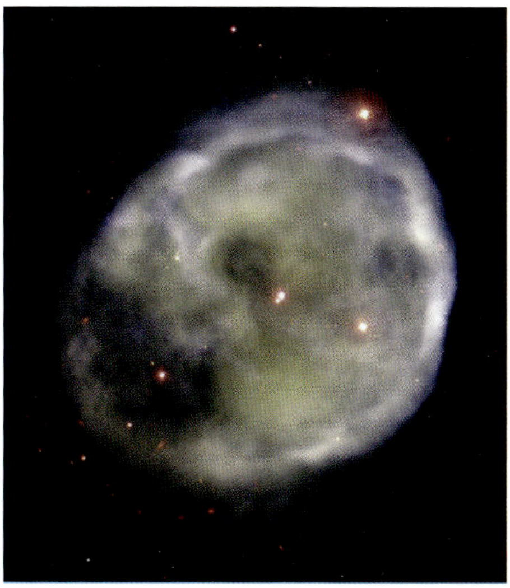

A superb Gemini South Observatory "GMOS" image of NGC 246 showing why it is sometimes referred to as the "Skull" nebula. The field is approximately 5 arcmin wide. Credit: Gemini Observatory/AURA/Travis Rector University of Alaska Anchorage.

Caldwell 57

NGC 6822
Galaxy in Sagittarius
Magnitude: 8.8
Size: 15′ × 13′
Mag/Sec: 23
RA: 19 h 44 min 54 s
Dec: –14° 48′
Circumpolar: below 76°S
Starhop: 6¼°N and 8½°E from Albaldah (π Sagittarii)
Best Months: June to August
Distance: 2.3 million light-years
Physical Size: ~10,000 light-years
Age: Billions of years
Best Visual Aperture: 10 cm and larger
Best Visual Filter: Deep sky filter may help
CCD/DSLR Advice: Deep, unfiltered exposures work best
Celestial Neighbors: Planetary neb. NGC 6818 lies 40′ NNW
Miscellaneous: Discovered by E.E. Barnard in 1884

It is hard to say anything complimentary about Caldwell 57's appearance in CCD images. In the deepest images, it resembles a luminous crisp packet being attacked by pink snails, or maybe a blood plasma bag spray-painted silvery-blue. Yes, we are back in the domain of the butt-ugly irregular dwarf galaxy, where we have already encountered NGC's 147, 185, and IC 1613 (Caldwell's 17, 18, and, "The Scarecrow," number 51). As this object barely climbs much higher than 20° altitude from northern latitudes we have a good reason for ignoring it.

If you imagine extending the handle of the famous Sagittarius teapot north-northeast toward the southern border of Aquila, NGC 6822 lies 3° below that border line. A planetary nebula, NGC 6818 can be found 40 arcmin north-northwest of the galaxy and is just as interesting. In images the dwarf galaxy does not resemble a traditional spiral, or "two fried eggs back to back," galaxy at all. It is rectangular, with the full width of the minor axis only becoming apparent if you stare hard enough. Its dimensions of 15′ × 13′ look more like 15′ × 9′. Around the galaxy edges there are an unusually large number of pink HII regions (the snails in the description) surrounding star-forming areas. Through a telescope the object is just a very subtle glowing pillar even in a 25-cm aperture under a crystal clear sky. With an estimated 250 million stars making up Caldwell 57 the galaxy is no more massive than the black holes at some galaxy cores and would look infinitely more dramatic if gravity had pulled them in to form a monster mother of all globular clusters rather than a pitiful sack of starlight.

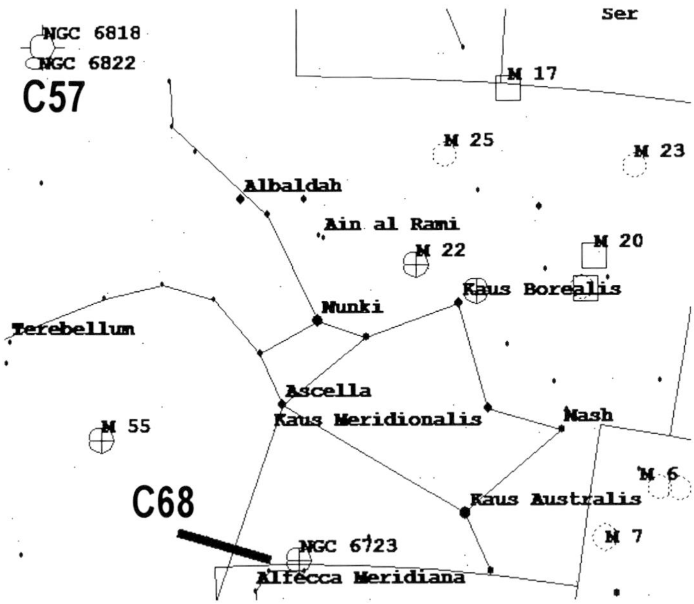

The location of Caldwell 57, in northeastern Sagittarius, is shown, along with Caldwell 68 just over the Corona Australis border.

This excellent image of the very low surface brightness galaxy NGC 6822 was taken by Tom Harrison from his Mano Prieto observatory in Texas USA, in October 2008, using a 12.5-in. *f*/9 RCOS Ritchey-Chrétien and Paramount ME, plus SBIG STL6303E CCD. LRGB exposures of 300:150:150:150 min.

Caldwell 58

NGC 2360
Open Cluster in Canis Major
Magnitude: ~7
Size: 14′
Mag/Sec: Not relevant
RA: 07 h 17 min 42 s
Dec: −15° 38′
Circumpolar: below 75°S
Starhop: 1°N and 8°E from Sirius (α Canis Majoris)
Best Months: December to February
Distance: 3,700 light-years
Physical Size: ~15 light-years
Age: ~2 billion years
Best Visual Aperture: Big binoculars and larger apertures
Best Visual Filter: No filter required
CCD/DSLR Advice: Easy object with any system
Celestial Neighbors: Many open clusters, esp. M41, 9°SW
Miscellaneous: One cluster in a cluster-packed region!
Chart: See Caldwell 54

The brightest star in the night sky is the steppingstone for reaching Caldwell 58. It lies just 1° north and 8° east of Sirius, the dog star, in Canis Major. Sirius has two stars roughly 5° either side of it, namely Murzim to the west (magnitude 2.0) and fainter Muliphen to the east (magnitude 4.1). Our seventh magnitude cluster lies 3½° due east of Muliphen. As mentioned when we dealt with Caldwell 54 the whole of southern Monoceros, northeastern Canis Major, and, especially, Puppis, is stuffed to the brim with open clusters, so Patrick was somewhat spoiled for choice when he made his selection. A fifth magnitude star, GSC 5965363, lies 20 arcmin west of this cluster and catches the eye if you do slide east from Muliphen to reach the field. Being superimposed on a dense portion of the Milky Way always makes it tricky to estimate where an open cluster actually ends, but the central 6 or 7 arcmin of NGC 2360 are very obvious.

The cluster contains about 90 stars packed into a 15 light-year diameter, with an average separation between stars of slightly more than 3 light-years. While this might not sound as dramatic as living in a globular cluster, the night sky from within the heart of this cluster would shine with a dozen Venus-brightness stars blazing down and dozens more of Jupiter's brightness, in other words, roughly magnitude −4 to −2. Through a telescope the brightest stars in NGC 2360 are of tenth magnitude, but the rest are of 11th to 12th. This is bright enough to make quite a visual impression through a 25–40 cm instrument, where most of the cluster's stars can be held steady without having to rely on averted vision. Although the cluster will not blow your mind it is well worth a look if you are in the region.

NGC 2360 from the Digitized Sky Survey. The field is 15 arcmin wide.

Caldwell 59

NGC 3242
Ghost of Jupiter Nebula in Hydra
Magnitude: 7.8
Size: 45″ × 35″
Mag/Sec: 15
RA: 10 h 24 min 46 s
Dec: −18° 39′
Circumpolar: below 72°S
Starhop: 10°S and 14°E from Alphard
Best Months: January to March, S. Hemisphere.
Distance: 1,400 light-years
Physical Size: ~2 light-years
Age: ~1,500 years (inner) to 100,000 years (outer)
Best Visual Aperture: 10 cm and larger apertures with high powers
Best Visual Filter: UHC or OIII filters work well
CCD/DSLR Advice: Good tracking at <1″/pixel desirable
Celestial Neighbors: Relatively barren surroundings
Miscellaneous: Some prefer "The Ghost of Uranus"
Chart: See Caldwell 53

Four objects ago we encountered the Saturn Nebula, so it was rather predictable that we would encounter one with Jupiter in the title at some point. We are back in Hydra for the "Ghost of Jupiter" Nebula, but we are now so far below the celestial equator that, for northern observers, at least, this planetary is badly affected by its maximum 20° altitude. NGC 3242 transits the meridian only 16 min later than brilliant Regulus in the Sickle of Leo, and that star can be used as a starting point to finding the planetary for us far northern residents. Travel a bit more than 24° due south from Regulus, and you will arrive at the magnitude 3.6 naked-eye star λ Hydrus. Head slightly more than 5° southeast from there, and a slightly fainter star, magnitude 3.8 μ Hydrus, can be found. From μ the planetary we seek is just 2° further south and a quarter of a degree west. Alternatively, if you can locate magnitude 2.0 Alphard (α Hydrus) moving 10° south, and 14° east will bag the planetary, too.

It is remarkable how all the planetary nebulae we seem to have come across so far have such a similar size. Apart from Caldwell 56 all the Caldwell planetary nebulae have been 20–45 arcsec across. But wait until we get to the Helix Nebula (Caldwell 63)!

The Ghost of Jupiter is another vaguely eye-shaped planetary nebula with dimensions similar to the apparent size of the planet it is nicknamed after, namely, 45″ × 35″. At seventh magnitude this is one of the brightest planetary nebulae, and it looks very star-like at low magnifications. Through a big telescope it looks surprisingly like an eye, and the Arizona amateur Steve Coe's nickname for it, "The CBS eye" (after the CBS TV station logo), is more appropriate than "Ghost of Jupiter." The planetary has a blue/green hue in large apertures, and the central star is variously reported as 11th to 12th magnitude, but, as with all central stars surrounded by nebulosity, it is surprisingly hard to see.

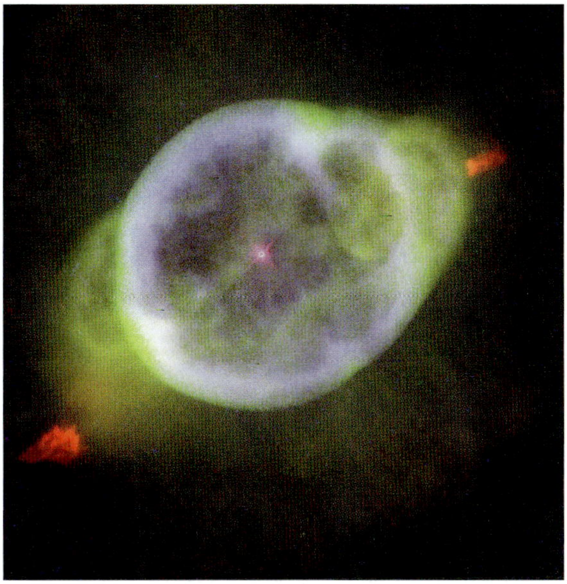

The planetary nebula NGC 3242 imaged in fine detail by the Hubble Space Telescope. Field of view approximately 50 arcsec. Credit: HST/NASA/ESA.

Caldwell 60 and Caldwell 61

NGC 4038 and NGC 4039
The Antennae in Corvus
Magnitudes: ~10.5 and 11.0
Size: 11′ × 6′ and 10′ × 4.5′
Mag/Sec: 23 and 24 respectively
RA: 12 h 01 min 53 s (4038) and 12 h 01 min 54 s (4039)
Dec: −18° 52′ (4038) and −18° 53′ (4039)
Circumpolar: below 72°S
Starhop: 1½°S and 3½°W from Gienah Ghurab (γ Corvi)
Best Months: February to April, S. Hemisphere.
Distance: 45 million light-years
Physical Size: ~140,000 light-years (each)
Age: Billions of years
Best Visual Aperture: 20 cm and larger apertures
Best Visual Filter: Deep Sky filter may help
CCD/DSLR Advice: Spectacular in mono or LRGB
Celestial Neighbors: Planetary neb. NGC 4361 lies 5°E
Miscellaneous: Also called "The Ringtail and Rattail galaxies"
Chart: See Caldwell 52

The two galaxies comprising the famous "Antennae" must surely be the best known examples of a massive cosmic collision. In Hubble images, ground-based professional images, and the best amateur images the scene is an awesome one. Two galaxies, each spanning at least a quarter of a million light-years of space, are taking part in a celestial punch-up! Of course, galaxies are mainly empty space, as the distance between stars is huge, except near the almost mandatory super-massive black hole at the center. In our own Milky Way, the average density is only 10^{-23} grams per cubic centimeter. But gravity is the dominant force over large distances and huge timescales, and when two galaxies comprising, perhaps, 50 billion stars attract one another it becomes a battle of tidal forces on an unimaginable scale.

The nearest bright star to the Antennae is magnitude 2.5 Gienah Gurab (γ Corvi). Most northern observers will not be familiar with Corvus the Crow, but Gienah Gurab is in the top right of the rectangular box comprising the Crow. Look roughly 20° east-southeast of brilliant Spica, and you will see the rectangle. From Gienah Gurab drop a degree and a third in Dec. and move 3 ° west, and you hit the Antennae. At magnitude 11 and 10.5 for these galaxies, you will definitely prefer a big aperture. But even with 30 cm at your disposal you will mainly see the bright center of NGC 4038 with the faint smudge of NGC 4039 making up the tail of a backwards comma, visually less than 5 arcmin in extent. NGC 4038 is a prolific supernova producer; four have been spotted occurring in the years 1921, 1974, 2004, and 2007. Of the two most recent events 2004gd was a Type 1b/c discovered by the amateur Berto Monard, and 2007sr was a valuable Type Ia discovered by Drake and his colleagues.

An amazing high power image of these interacting galaxies by the Hubble Space Telescope. Image: NASA, ESA, and the Hubble Heritage (STScI/AURA) ESA-Hubble Collaboration HST/ACS

The awesome interacting galaxies NGC 4038 and NGC 4039, the "Antennae," imaged expertly by Robert Gendler using 12 h of LRGB exposures obtained remotely with a 14.5″ (37 cm) f/9 Ritchey-Chrétien with RCOS field flattener, based in Western Australia. Paramount ME mount and SBIG STL-11000XM CCD. © Robert Gendler.

Caldwell 62

NGC 247
Milkweed Seed Galaxy in Cetus
Magnitude: 9.2
Size: 22′ × 7′
Mag/Sec: 23
RA: 00 h 47 min 06 s
Dec: –20° 46′
Circumpolar: below 70°S
Starhop: 2¾°S and ¾°E from Diphda (ß Ceti)
Best Months: September to November, S. Hemisphere.
Distance: 7 million light-years
Physical Size: ~45,000 light-years
Age: Billions of years
Best Visual Aperture: 15 cm and larger apertures
Best Visual Filter: Deep sky filter may help
CCD/DSLR Advice: Good in mono or LRGB
Celestial Neighbors: C65 (NGC 253) lies 4½°S
Miscellaneous: No NGC 247 supernovae discovered yet
Chart: See Caldwell 56

As we move further south than Dec. –20 we start to encounter some really big and impressive deep sky objects that, while they are virtually inaccessible from some northern locations, knock some of Patrick's northern hemisphere offerings into the proverbial cocked hat. Why should this be? Well, the Caldwell catalog deliberately excludes the bright Messier objects, and Messier observed from Paris. At a latitude of almost 49° north anything below Dec. –20 would have been considerably dimmed and blurred by the atmosphere and only visible for short periods; anything below Dec. –41 would never rise at all. The most southerly galaxy in Messier's catalog is the big and bright galaxy M83, at Dec. –29, discovered by his countryman Lacaille on a trip to South Africa in 1751–1752. The most southerly Messier object is the third magnitude open cluster M7 in Scorpius, at (almost) Dec. –35. Essentially, the further south we go below about Dec. –20 the more we start bagging superb objects that Messier could never have seen. NGC 247 is a fantastic galaxy for CCD imagers and not bad for visual observers, either. Although not a physically big galaxy its distance of 7 million light-years makes, it look huge and a perfect size for the fields of view of many amateur CCD imaging systems. At more than a third of a degree across, it can nicely span the chip. If there is one negative aspect it is simply the galaxy's rather low surface brightness; a galaxy with dimensions of 22′ × 7′ should be a bit brighter than magnitude 9. This, of course, is less of an issue with long CCD exposures. There is just a hint of the irregular "dwarf"-type appearance we encountered with Caldwell's 17, 18, and 51, but this galaxy almost 3° south and almost a degree east from Diphda (ß Ceti) has made the effort to look good. But its brother, NGC 253, almost 5° to the south, is even better; wait till we get to Caldwell 65!

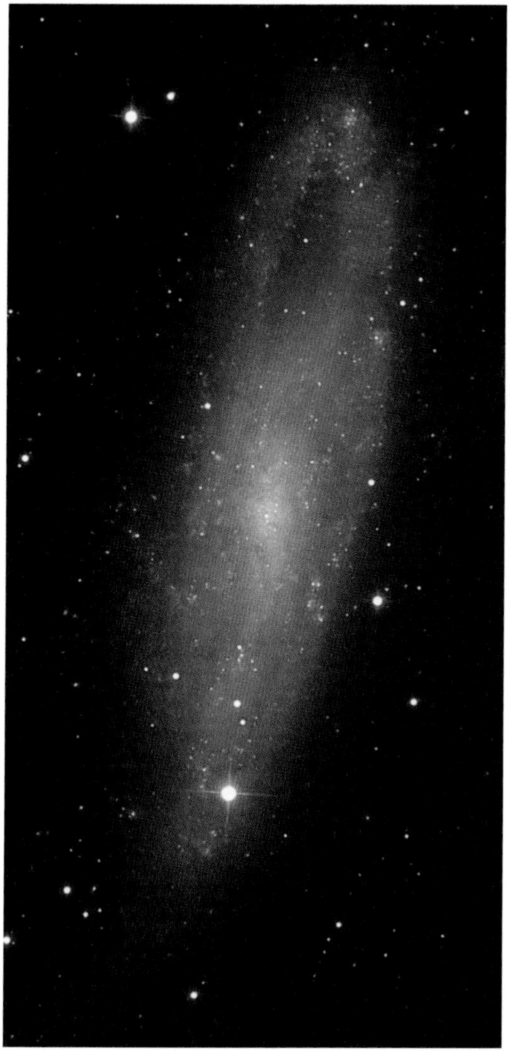

This excellent image of NGC 247 was taken by Josef Pöpsel and Stefan Binnewies on 2006 May 4 from the Capella Observatory, then located at Amani Lodge, Kupferberg near Windhoek, Namibia. A Ganymed 60 cm-Hypergraph was used at primary focus, *f*/3, *f* = 180 cm. SBIG STL-11000M CCD. RGB 4 × 300 s: 3 × 300 s: 3 × 300 s. Image by kind permission of Stefan Binnewies and Josef Pöpsel.

Caldwell 63

NGC 7293
Helix Nebula in Aquarius
Magnitude: 7.3
Size: 12′ × 10′
Mag/Sec: 20–21
RA: 22 h 29 min 38 s
Dec: –20° 50′
Circumpolar: below 70°S
Starhop: 9¼°N and 6½°W from Fomalhaut (α PsA)
Best Months: July to October, S. Hemisphere.
Distance: 520 light-years
Physical Size: ~22 light-months
Age: ~10,000 years
Best Visual Aperture: Try 10 cm and larger at ~60×
Best Visual Filter: OIII best; UHC good too
CCD/DSLR Advice: Colorful, so LRGB is recommended
Celestial Neighbors: Relatively barren surroundings
Miscellaneous: Disc. by G.F.J.A. (Arthur) von Auwers ~ 1860?

All but one of the Caldwell planetary nebulae we have encountered so far have been tiny – less than an arcminute across. The Helix, in Aquarius, is different; by planetary nebula standards it is HUGE, a whopping 10 by 12 arcmin. The reason for its size is its distance. Most planetary nebulae are thousands of light-years away, but the Helix is only about 520. It lies slightly more than 10° northwest of the first magnitude star Fomalhaut, and magnitude 5.2 upsilon Aquarii lies a degree to the east but, to be honest, this planetary is hard to miss in big binoculars and finder telescopes!

Being one fifth of a degree across a 60° apparent field eyepiece at 300× will cause it to fill the field, but you may find this is a backward step. Most people's night vision rods work best when an object is placed 8–16° off center. Take the average as 12° and match this to the Helix' radius of a fifth of a degree and 60×; purely from a theoretical perspective, this seems a good magnification to use. With, say, a 300 mm scope this gives a 5 mm exit pupil, which even senior irises can accommodate. Some say the Helix is of low surface brightness, but if it is as bright as estimated by Stephen O'Meara, namely sixth magnitude, it glows (ignoring the hole) at around magnitude 20/sq. arcsec. Yes, this is faint for a planetary nebula, as some are 100 times brighter per unit area, but if the Helix was a galaxy it would be one of the brightest on a unit area basis. But why struggle without some help? The use of a UHC or, especially, an OIII filter will dramatically darken the sky background and really make the Helix a stunning sight in any telescope. With such a nearby planetary nebula one might expect the central star to be bright, but it is only magnitude 13.5 and does not stand out from the crowd. However, infrared Spitzer space telescope images of the star and its dusty environs show it as a beacon at the heart of the Helix.

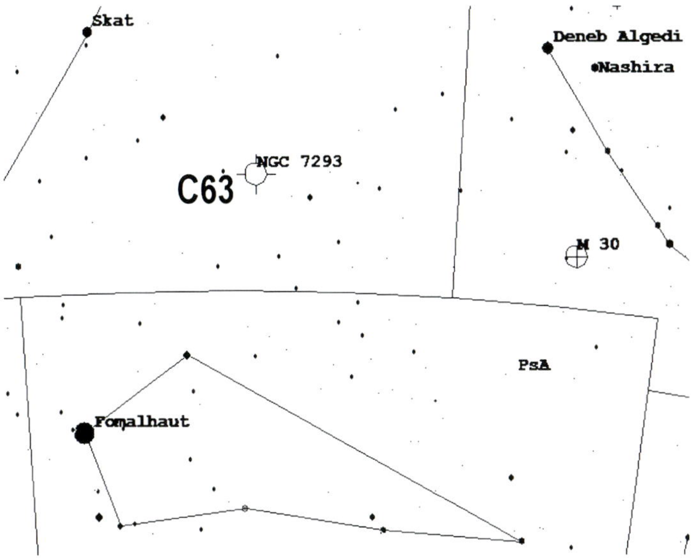

The location of Caldwell 63 in southwestern Aquarius, northwest of the bright star Fomalhaut, is shown.

Remarkably this spectacular image of the Helix Nebula was taken by an amateur astronomer; once again this is an image by Tom Harrison at his Texas "Mano Prieto Observatory." 12.5-in. *f*/9 RCOS Ritchey-Chrétien plus Paramount ME mounting and SBIG STL6303E CCD. Images exposed in November 2007.

Caldwell 64

NGC 2362
Tau Canis Majoris Cluster
Magnitude: 4
Size: 6′
Mag/Sec: Not relevant
RA: 07 h 18 min 42 s
Dec: −24° 57′
Circumpolar: below 66°S
Starhop: 8¼°S and 8°E from Sirius (α Canis Majoris)
Best Months: December to March, S. Hemisphere.
Distance: 5,000 light-years
Physical Size: ~9 light-years
Age: ~25 million years
Best Visual Filter: 10 cm is sufficient
Best Visual Aperture: Filter not required
CCD/DSLR Advice: Easily imaged with modest telescopes
Celestial Neighbors: Open cluster NGC 2354 lies 80′ to SW
Miscellaneous: Disc. by Giovanni Batista Hodierna ~ 1650?
Chart: See Caldwell 54

The reader may have noticed that viewing from the UK is often mentioned. In the UK it is bitterly cold at night from October through to May, and then just as the night-time temperatures get reasonable we get permanent twilight, writing June and July off. Only August and September offer warm nights and dark skies. Add to this the fact that anything further south than Dec −30 is, in practice, impossible, and as we get to Caldwell numbers in the high sixties Brits are stuggling to see them. Caldwell 64 is really the last hope binocular Caldwell object for England, but, fortunately, it is dead easy to find. Everyone can find Sirius and, 10° to its south-southeast, the star Wezen, at magnitude 1.8. At best, Wezen (δ Canis Majoris) rises 11° above the south horizon, but it is bright enough for this not to be an issue. Do not mistake Wezen for Adara (ε Canis Majoris), though, 3° southeast from Wezen and slightly brighter. From Wezen, the Tau Canis Majoris Cluster is 3° to the north-east. There is a slight problem, though. On your way from Wezen to Caldwell 64, you will sweep over the sixth magnitude open cluster NGC 2354. Do not mistake it for the Caldwell object! Our target, the Tau Canis Majoris Cluster, is sometimes called the "Northern Jewel Box" (see Caldwell 94 for the real article). The "Jewels" are 20 stars of roughly ninth and tenth magnitude, two of eighth magnitude, and the crown jewel, Tau Canis Majoris itself, of magnitude 4.4. Remember the "Blinking Planetary," Caldwell 15, where you see the central star with direct vision and the nebula with averted vision? This cluster's bright star can play funny tricks, too. Although not visible to amateur telescopes Tau Canis Majoris has four companion stars; it is a quintuple star system. Although not spectacular, Caldwell 64 really does resemble a sparkling box of jewels in any modest aperture telescope.

NGC 2362 from the Digitized Sky Survey. The field is 10 arcmin wide.

Caldwell 65

NGC 253
Silver Coin Galaxy in Sculptor
Magnitude: 7.6
Size: 26′ × 6′
Mag/Sec: 21
RA: 00 h 47 min 36 s
Dec: –25° 17′
Circumpolar: below 65°S
Starhop: 7¼°S and ¾°E from Diphda (ß Ceti)
Best Months: September to November, S. Hemisphere.
Distance: 10 million light-years
Physical Size: ~75,000 light-years
Age: Billions of years
Best Visual Aperture: A fine sight in 10 cm and larger apertures
Best Visual Filter: Deep sky filter may help
CCD/DSLR Advice: Awesome object in mono or color
Celestial Neighbors: C62 & Globular NGC 288 (100′ SE)
Miscellaneous: Host to Supernova 1940E disc. by Fritz Zwicky
Chart: See Caldwell 56

It is tempting to say that Caldwell 65 is this author's favorite Caldwell object. It is also the furthest south object in Patrick's catalog that is generally seeable from a UK site. Even at a transiting altitude of about 12° at midnight at the start of October, it looks impressive in a 35-cm aperture Celestron 14, sky transparency permitting. Not many galaxies are as long as NGC 253, although NGC 247, which we encountered three objects ago, comes close, and NGC 55 (C72) is even longer. However, NGC 253 has a much higher surface brightness, which makes all the difference. NGC 247 can sometimes be spotted with the same telescope on the same night, and it is a ghost compared with NGC 253. Despite the glancing angle at which we view NGC 253, there is still plenty of detail visible from southerly latitudes, which look spectacular in CCD images. Two major spiral arms are easy to make out, and the detail in them, the dust lanes, and the central bar is awesome.

Once again the bright star Diphda is the key to locating the "Silver Coin" galaxy and, frankly, only a madman would look at this galaxy in isolation without comparing it to the aforementioned NGC 247 4½ degrees to the north. Both galaxies are, gravitationally, members of the Sculptor group, even if NGC 247 is over the border in Cetus. From Diphda (ß Ceti) track southwards by 7¼°, then east by ¾ of a degree, and NGC 253 will hit you. Something else might hit you, too, if you overshoot the field. The globular cluster NGC 288 lies less than 2° to the southeast – a fine eighth magnitude deep sky object in its own right. Maybe the only problem with the Silver Coin Galaxy (and it does resemble a coin seen at a low angle) is imaging its 26 arcmin length; with a 2 m focal length the galaxy will span 15 mm on the chip. Be prepared to have to join two images together!

The stunning Sculptor galaxy NGC 253, imaged by Tom Harrison at his Texas "Mano Prieto Observatory." 12.5-in. *f*/9 RCOS Ritchey-Chrétien plus Paramount ME mounting and SBIG STL6303E CCD. Images exposed in November 2007. LRGB 360:120:120:120 minutes.

Caldwell 66

NGC 5694
Globular Cluster in Hydra
Magnitude: 10.2
Size: 4.3′
Mag/Sec: 22
RA: 14 h 39 min 37 s
Dec: –26° 32′
Circumpolar: below 64°S
Starhop: 10½°S and 2½°W from Zuben Elgenubi (α Librae)
Best Months: April to July, S. Hemisphere
Distance: 110,000 light-years
Physical Size: ~140 light-years
Age: ~12 billion years
Best Visual Aperture: 20 cm and larger
Best Visual Filter: No filter advantage
CCD/DSLR Advice: Easy to image in mono or color
Celestial Neighbors: Relatively barren surroundings
Miscellaneous: Southern hemisphere equivalent of C42

After the excitement of Caldwell 65 it is back to something rather duller – one of those very distant globulars that would look a lot healthier if only they were closer. Still, there is some satisfaction in chasing these critters down. NGC 5694 lies 10½° south and 2½° west of the bright star Zuben Elgenubi in Libra, perhaps best known because it sounds like a Star Wars character who might have taught his apprentices well in mastering the power of the Force. A distinctive arc of five fifth magnitude stars curving round from 4 Librae (star 1) to 57 Hydrus (star 5) lies roughly 2° to the east and northeast of the globular. Another pointer to this faint globular is the third magnitude star Brachium (σ Librae), which lies less than 5° to the east-northeast of the globular on the opposite side of the aforementioned five-star arc. Closer in, with a 40-arcmin-wide field centered on the globular, two other stars become very conspicuous. A seventh magnitude star lies 10 arcmin south of the globular, and another seventh magnitude star lies 20 arcmin east-northeast of the object. Through a telescope two tenth magnitude stars trail off the southwestern edge of the globular and are obvious in the image shown here. Any observer worth his or her salt, while in this region, will not miss the opportunity to digress 12° west-southwest to glimpse the fabulous supernova producing galaxy M83, also in Hydra. So, what of Caldwell 66 itself? Well, it is a tiny globular and easy to mistake for a star at low powers. If you live south of the equator and can bring to bear a powerful telescope and high magnifications, under steady seeing, onto this 4-arc-min fuzz you may be able to resolve some interior detail. Otherwise it is just another deep sky object to tick off the list.

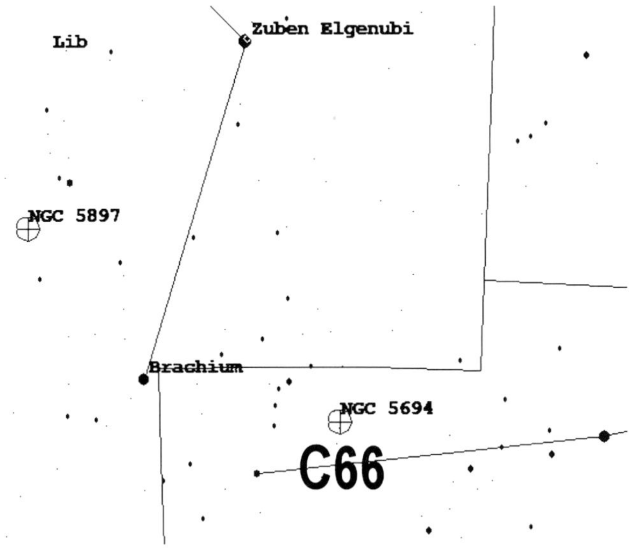

The chart shows the location of Caldwell 66 in Hydra, just south of the Libra border and the distinctively named star Zuben Elgenubi.

NGC 5694 from the Digitized Sky Survey. The field is 10 arcmin wide.

Caldwell 67

NGC 1097
Galaxy in Fornax
Magnitude: 9.2
Size: 10′ × 6′
Mag/Sec: 22
RA: 02 h 46 min 18 s
Dec: –30° 16′
Circumpolar: below 60°S
Starhop: 1°S and 6°W from Fornacis (α For)
Best Months: September to January, S. Hemisphere.
Distance: 45 million light-years
Physical Size: ~130,000 light-years
Age: Billions of years
Best Visual Aperture: 15 cm and larger
Best Visual Filter: Deep sky filter may help
CCD/DSLR Advice: Great LRGB target
Celestial Neighbors: Many fainter galaxies nearby
Miscellaneous: Prime supernova hunters galaxy

NGC 1097 is another magnificent southern hemisphere galaxy, although its true beauty is only revealed in deep images. We view it only 30° from being face-on and so can enjoy the galaxy's two dramatic, sweeping spiral arms to the full. The galaxy's center appears as an elliptical bar shining far more brightly than the spiral arms. This is not too surprising, as NGC 1097 is a Seyfert galaxy, a particularly active galaxy with an energetic core, no doubt fueled by a super-massive black hole at the center. The whole of the barred center looks a bit like a luminous melon in the finest amateur images. The galaxy experienced three Type II supernovae in the 11-year span between 1992 and 2003. Supernova 1992bd was discovered by Chris Smith and Lisa Wells at Kitt Peak Observatory on October 12 of that year. Seven years later, in November 1999, the Japanese amateur Aoki discovered another supernova 2½ arcmin south of NGC 1097's core. Then, in January 2003, the legendary visual supernova discoverer Robert Evans discovered the third supernova almost 3 arcmin north of the galaxy's core.

This is definitely a galaxy worth checking on a regular basis. Three supernovae in such a short period may seem a bit of a fluke, but NGC 1097 may contain as many as 200 billion stars, maybe more than our own Milky Way. The more stars, the more chance of a supernova in a given time period. Perched just on the northwest tip of NGC 1097 is the 13th magnitude galaxy NGC 1097A. Finding NGC 1097 is a bit painful, as there are no bright stars anywhere near the region. The brightest star in Fornax (do not get too excited) is magnitude 3.9 Fornacis. From there it is a hop just less than 6° west and just more than a degree south. Visually, the spiral arms are a challenge unless you have a big telescope, pristine skies, and a lot of experience. The same applies to seeing NGC 1097A.

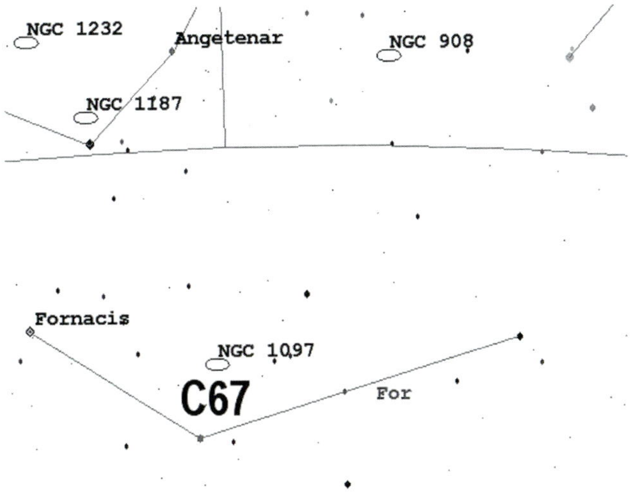

NGC 1097 (C 67) sits just above the shallow "V" made by the brightest stars in Fornax in the *lower left* of the chart.

This excellent image of NGC 1097 was taken by Dietmar Böcker, Josef Pöpsel and Stefan Binnewies on 2004 October 16 from the Capella Observatory, then located at Amani Lodge, Kupferberg near Windhoek, Namibia. A Ganymed 60 cm-Hypergraph was used at primary focus, *f*/3, $f = 180$ cm. SBIG ST10 XME CCD. LRGB. L: 15×180 s. 1×1 bin; R,G,B: 5×180 s. each, 2×2 bin. Image by kind permission of Stefan Binnewies and Josef Pöpsel.

Caldwell 68

NGC 6729
Nebula in Corona Australis
Magnitude: 9.5
Size: ~80″
Mag/Sec: 17
RA: 19 h 01 min 54 s
Dec: –36° 57″
Circumpolar: below 54°
Starhop: 2½°S and 7¾° E of Kaus Australis (ε Sgr)
Best Months: May to August, S. Hemisphere.
Distance: 420 light-years
Physical Size: ~2 light- months
Age: ~200,000 years?
Best Visual Aperture: 15 cm and larger
Best Visual Filter: Deep sky filter may help
CCD/DSLR Advice: Responds well to unfiltered imaging
Celestial Neighbors: NGC 6723, NGC 6726/7 and IC 4812
Miscellaneous: Disc. by Albert Marth in the 1860s
Chart: See Caldwell 57

Corona Australis is a tiny constellation; the ninth smallest in area out of the 88 modern constellations. The feature Patrick allocated the number 68 to lies just within this constellation. Move a smidgen further north, and you are immediately in southern Sagittarius. Apart from the confusing proximity to the constellation boundary there are many nearby distractions, too, and they are bigger, brighter and more appealing than NGC 6729. To get to Caldwell 68 one strategy might be to travel 2½° south and almost 8° east of Kaus Australis, the first magnitude star at the southwestern tip of the Sagittarius' teapot base. En route from that star you may well encounter the fine seventh magnitude globular cluster NGC 6723, which was excluded from the Caldwell listing even if Caldwell 66, a globular twenty times fainter, was included; do feel free to spend more time on the globular!

The distractions do not end there, though. The faint nebula NGC 6726/6727 is a tenth of a degree to the northwest of the object we seek and is also a brighter object. Indeed, it is probably worth regarding NGC 6726/6727 and 6729 as one associated eyepiece object. The former duo consists of an 8 arcmin fuzz surrounding the magnitude 7.3 star GSC 7421976 and the variable star TY Coronae Australis, which varies between magnitude 8.8 and 12.6. Caldwell 68 itself lies roughly 5 min of arc to the southeast of the bigger NGC 6726/6727 fuzz. At high power it looks a bit like Hubble's Variable Nebula in Monoceros (Caldwell 46), and this one varies in appearance, too. The "comet" trails away to the southeast from its illuminator, R Coronae Borealis. On its own Caldwell 68 is rather dull, but the nearby globular and other nebulosity make the trip here worthwhile, as does the 6th magnitude double star ¼° to the south, and the associated nebulosity IC 4812.

The wide field image (oriented north-northeast up) shows two significant areas of nebulosity on the left (NGC 6726-7 upper, IC 4812 lower) and a globular cluster (NGC 6723) on the right hand side. The much smaller comet-like nebulosity just to the left of NGC 6726-7 is Caldwell 68, or NGC 6729. The close-up image (north up) shows the NGC 6726-7/Caldwell 68 region in more detail. These superb images are by Volker Wendel and Bernd Flach-Wilken. Wide field shot: a 5-in. (12.5 cm) *f*/6-Astrophysics Refractor and SBIG ST10XME. LRGB ($L = 10 \times 300$ s; $R = 5 \times 300$ s; $G = 3 \times 300$ s; B=4x600s; RGB binned 2×2 combined with a LRGB set of 30-s exposures for the brightest parts). Narrower field shot: 16-in. (40 cm) f/8 Hypergraph in Namibia with an SBIG ST10XME with Astronomik-Filterset. LRGB ($L = 6 \times 300$ s; $R = 3 \times 300$ s; $G = 3 \times 300$ s; $B = 3 \times 300$ s; RGB binned 2×2). Copyright © 2004 Volker Wendel and Bernd Flach-Wilken.

Caldwell 69

NGC 6302
Bug Nebula in Scorpius
Magnitude: 9.5
Size: 85″ × 25″
Mag/Sec: 18
RA: 17 h 13 min 44 s
Dec: –37° 06′
Circumpolar: below 53°S
Starhop: 4° due W from Shaula (λ Scorpius)
Best Months: May to August, S. Hemisphere.
Distance: 5,000 light-years
Physical Size: ~2 light-years
Age: ~10,000 years
Best Visual Aperture: 15 cm and larger with high powers
Best Visual Filter: OIII or UHC
CCD/DSLR Advice: Small object, best imaged at <1″/pixel
Celestial Neighbors: Large open cluster NGC 6821 to the SW
Miscellaneous: Disc. by E.E. Barnard in 1883
Chart: See Caldwell 78

It seems appropriate that a nebula shaped vaguely like a winged insect should be located in the constellation named the Scorpion. Apparently this is the most southerly object that can be imaged from the Kitt Peak Observatory Visitors Center in Arizona (latitude 37° 57′ N), where it transits at a maximum altitude of 15°. To find the field just head for the Scorpion's sting star Shaula (magnitude 1.6) and simply slide 4° due west. The Bug seems to be a much neglected object among visual and CCD observers alike. One would surely expect a veritable swarm of observations of an object resembling an insect that was 2 light-years in length! However, it only really seems to have attracted significant attention following Patrick's listing of it in his catalog and a spectacular image of it being captured by the Hubble Space Telescope. Undoubtedly, as we move well into the deep south of the sky such neglected objects will emerge because the majority of deep sky observers live in the far northern hemisphere.

In the best amateur CCD images the central part of the Bug Nebula looks like a pink or red butterfly with a symmetrical gossamer wing structure to the east and west. The edges of the wings are dramatically sharp and almost identical. Outside the central region it looks like the butterfly is perched on a pink firework, with the stick and some residual smouldering sticking out to the west. Images of the planetary nebula taken in hazy skies will show less of the butterfly's wings and more of the firework itself. Visually, even in a large backyard telescope, the Bug just looks like a long gray streak. That awesome 2004 Hubble image of the center of the bug (where the wings join) showed that the Bug's outer regions of dust contain hydrocarbons, carbonates such as calcite, as well as water-ice and iron. The central star is hidden from view by the dust but is thought to be at the astounding temperature of 250,000°C!

NGC 6302, the bug nebula, imaged by Adam Block at the Kitt Peak Observatory Visitor Center with a 16-in. (0.4 m) f/10 Schmidt-Cassegrain and SBIG ST8E CCD camera with color filter wheel. LRGB 45:10:10:20 minutes. (R, G, and B binned 2 × 2). Credit: Adam Block/NOAO/AURA/NSF.

A close up of The Bug's centre imaged by the Hubble Space Telescope. Slicing across a bright cavity of ionized gas, the dust torus surrounding the central star is in the upper right corner, almost edge-on to the line-of-sight. Credit: A. Zijlstra (UMIST) et al., ESA, NASA

Caldwell 70

NGC 300
Southern Pinwheel Galaxy in Sculptor
Magnitude: 8.1
Size: 20′ × 15′
Mag/Sec: 23
RA: 00 h 54 min 54 s
Dec: –37° 41′
Circumpolar: below 53°S
Starhop: 4½°N and 5½°E from Ankaa (α Phe)
Best Months: August to December, S. Hemisphere.
Distance: 4 million light-years
Physical Size: ~25,000 light-years
Age: Billions of years
Best Visual Aperture: 15 cm and larger is recommended
Best Visual Filter: Deep sky filter may help
CCD/DSLR Advice: LRGB delivers great results
Celestial Neighbors: C 72 (NGC 55) 8° further W
Miscellaneous: Still awaiting a supernova discovery
Chart: See Caldwell 56

There seems to be a line of Caldwell galaxies stretching south from the bright star Diphda (β Ceti). We have already ticked off the first two (NGC 247 and 253), and NGC 300 is the third specimen. NGC 55 is now only 40 min of right ascension west, and, logically, any keen southern hemisphere galaxy chaser would have a look at that galaxy on the same night. Frustratingly you will have to temporarily ignore the Caldwell 71 text in this book to get to NGC 55, as its Caldwell number is 72. NGC 300 lies 4½° north and 5½° east from the second magnitude star Ankaa (α Phe), the brightest star in the constellation of the Phoenix. The galaxy is sometimes referred to as the Southern Pinwheel for a reason; it does bear comparison with the northern Pinwheel, M33, in Triangulum. M33 is at least 30% nearer than NGC 300, and we see the Messier galaxy almost face-on, compared to our 45° view of the Caldwell galaxy. Nevertheless, when you see deep amateur images of M33 and NGC 300 side by side the similar structure of the two galaxies is quite striking. Both galaxies have a comparably low average surface brightness of around magnitude 23/sq. arcsec, so crystal clear skies at a dark site will give the best visual view.

If you live around latitude 35S to 40S the galaxy can pass overhead, too, improving prospects. A low magnification of around a quarter of the telescope aperture in millimeters will probably be optimum, but, even then, do not expect to see more than two-thirds of NGC 300's diameter. The outer edge of the spiral arms are only fractionally above the sky background, even at a desert location. The galaxy is a perfect size for big CCD detectors on a large Schmidt-Cassegrain at f/6.3 or so. Deep LRGB or RGB exposures will bring out the light blue arms peppered with pink HII regions.

This superb image of NGC 300 was obtained remotely by Robert Gendler using 11 h of LRGB exposures obtained remotely with a 14.5″ (37 cm) *f*/9 RCOS Ritchey-Chrétien with field flattener, based in Western Australia. Paramount ME mount and SBIG STL-11000XM CCD. © Robert Gendler.

Caldwell 71

NGC 2477
Open Cluster in Puppis
Magnitude: 5.8
Size: 20′
Mag/Sec: Not relevant
RA: 07 h 52 min 12 s
Dec: –38° 32′
Circumpolar: below 52°S
Starhop: 1½°N and 2¼°W from Naos (ζ Puppis)
Best Months: December to March, S. Hemisphere.
Distance: 3,700 light-years
Physical Size: ~20 light-years
Age: ~700 million years
Best Visual Aperture: 10 cm and larger
Best Visual Filter: No filter required
CCD/DSLR Advice: Easy imaging target
Celestial Neighbors: Brighter NGC 2451 is 90′ WNW
Miscellaneous: Disc. by de Lacaille from southern Africa ~ 1752

As we have headed ever southward down through the northern hemisphere winter Milky Way, passing through Monoceros and Canis Major, skirting Sirius and ploughing down into the southern hemisphere summer Milky Way we have traveled past dozens of open clusters but only glanced at three examples. These were Caldwell 54 in Monoceros, Caldwell 58 in Canis Major, and Caldwell 64, also in Canis Major. We have now dropped a further 13½° in declination along that river of stars (and moved half an hour of right ascension to the east, ignoring two dozen other open clusters) to arrive at Patrick's seventy-first choice for the Caldwell catalog. The odd thing about Caldwell 71 is that visually, or photographically, at first glance you do a double-take. Is this an open cluster, or a globular cluster? It seems to be neither one thing nor the other! This puzzling appearance is probably a mixture of stellar coincidences and characteristics.

First, unlike so many open clusters, there are no eye-catchingly bright stars in the cluster, or in front of the cluster. Second, the vast majority of the easily observable stars in the cluster (more than one hundred) are the same brightness, namely between magnitude 13 and 14. There are 2,000 much fainter stars within the cluster, but these are not visible to the eye, even in a large telescope. Third, the open cluster is remarkably spherical in appearance, with a gradual increase in star density toward the middle, just like a globular cluster. Although Patrick chose NGC 2477 as his favorite cluster in this part of the sky another dramatic open cluster lies just next door. NGC 2451 is bigger and brighter than NGC 2477 and has a magnitude 3.5 star at its heart as well as a dozen bright stars of fifth to eighth magnitude; a complete contrast to its neighbor. NGC 2477 is just 1½° north and 2¼° west from the second magnitude star Naos (ζ Puppis). NGC 2451 is a mere 1½° further to the west-northwest.

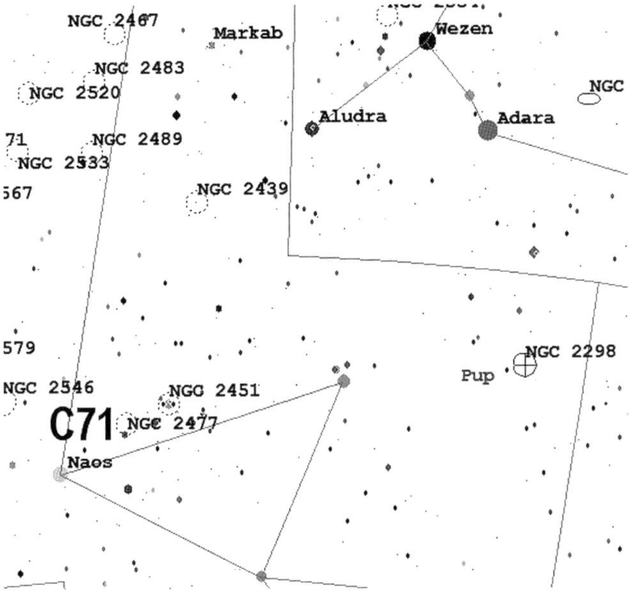

NGC 2477 lives in Puppis, not far from the second magnitude star zeta Puppis (Naos) as shown in the chart.

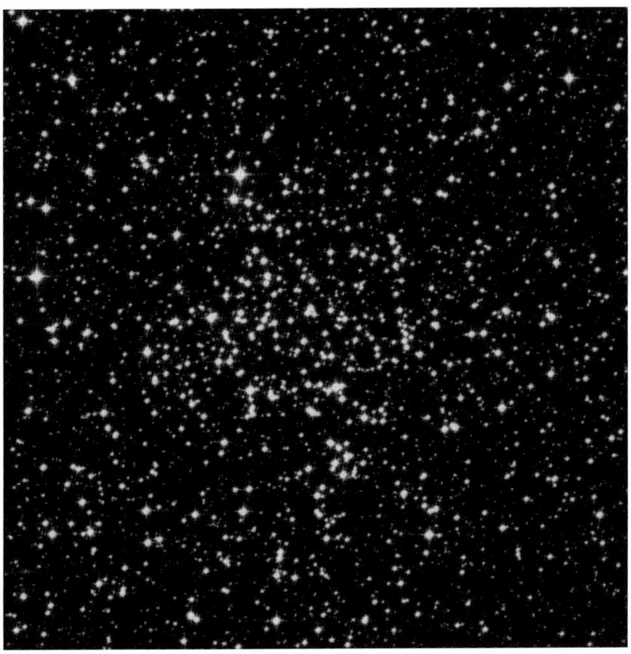

NGC 2477 from the Digitized Sky Survey. The field is 15 arcmin wide.

Caldwell 72

NGC 55
String of Pearls Galaxy in Sculptor
Magnitude: 8.1
Size: 34′ × 6.5′
Mag/Sec: 22–23
RA: 00 h 15 min 06 s
Dec: −39° 13′
Circumpolar: below 51°S
Starhop: 3°N and 2°W from Ankaa (α Phe)
Best Months: August to December, S. Hemisphere.
Distance: 4 million light-years
Physical Size: ~40,000 light-years
Age: Billions of years
Best Visual Aperture: 15 cm and larger
Best Visual Filter: Deep sky filter may help
CCD/DSLR Advice: Unfiltered imaging works well
Celestial Neighbors: C 70 (NGC 300) 8° further E
Miscellaneous: One of the sky's longest galaxies
Chart: See Caldwell 56

NGC 55 was mentioned just before the last Caldwell object since it lies next door to that other galaxy in Sculptor NGC 300 and can also be hopped to from the brightest star in the constellation of the Phoenix, namely Ankaa. The hop is 3° north and 2° west. NGC 55 could not be closer to the southern border of Sculptor; northern Phoenix is just a hair's breadth to the south. Although appearing significantly longer than Caldwell 65/NGC 253, the much nearer NGC 55 is nowhere near as spectacular in deep CCD images. Both galaxies are close to being edge-on to us, but NGC 55 lacks the spiral arms and splendid dust lanes of NGC 253. Indeed, the galaxy is strikingly asymmetric, with the eastern side being much longer but significantly fainter than the western side. Indeed the region just to the east of the brighter galaxy center seems to disappear in mediocre images or where the sky background is too bright. At a declination of −39° we are now forever struggling to get a good view of this galaxy from latitudes further north than +30°. As soon as an object drops to 20° in altitude and lower, atmospheric extinction rapidly takes its toll. Therefore, the majority of deep sky observers, living in northern Europe and North America, will not be familiar with NGC 55 or any of the 36 other remaining Caldwell objects. Despite the faint easterly extension of the galaxy the inner core of NGC 55 (offset to the west by about 3 arcmin from the geometric center) does have a detailed and bright appearance. Bearing in mind the extreme length of NGC 55 and the fact most of its brightness is contained in the core, it actually resembles an object with dimensions of roughly 15′ × 4′ in a telescope eyepiece, rather than its true length of 34′ × 6.5′. Nevertheless, this is still a very rewarding deep sky object to hunt down if you live far enough to the south.

The edge-on galaxy NGC 55, portrayed in another superb, remotely imaged shot, by Robert Gendler. Eight hours of LRGB exposures with a 14.5-in. (37 cm) *f*/9 RCOS Ritchey-Chrétien with field flattener, based in Western Australia. Paramount ME mount and SBIG STL-11000XM CCD. © Robert Gendler.

Caldwell 73

NGC 1851
Globular Cluster in Columba
Magnitude: 7.1
Size: 12′
Mag/Sec: 21
RA: 05 h 14 min 06 s
Dec: –40° 03′
Circumpolar: below 50°S
Starhop: 6°S and 5°W from Phaet (α Columbae)
Best Months: November to February, S. Hemisphere.
Distance: 40,000 light-years
Physical Size: ~140 light-years
Age: ~10 billion years
Best Visual Aperture: 10 cm and larger
Best Visual Filter: No filter advantage
CCD/DSLR Advice: Easy imaging target
Celestial Neighbors: Galaxies NGC 1792 and 1808 lie 3°NW
Miscellaneous: Worth comparing to M79

Roughly an hour and a half west of a line joining the brilliant stars Sirius and Canopus and bang on the line of declination –40 you can find the globular cluster NGC 1851. You can precisely center it by locating the magnitude 2.6 star Phaet and sliding 6° south and 5° west.

Globular cluster fanatics at equatorial or southern hemisphere locations may wish to compare the well known globular M 79 with Caldwell 73 on the same night. The Messier object is just 15° further north and 10 min of right ascension further east but is smaller and dimmer than this Caldwell offering. At a distance of only 40,000 light-years a couple of dozen of the stars in NGC 1851 are between thirteenth and fifteenth magnitude so, in theory, with a big enough amateur scope you will be able to hold these steady and see the object as a globular and not just a globular-resembling fuzz. With the integrating power of a CCD camera on even modest telescopes the bulk of the globular's mainly sixteenth and seventeenth magnitude stars shine as distinct points. John Herschel described NGC 1851 as "a superb globular cluster" and "very suddenly much brighter in the middle to a blaze or nucleus of light." Two bright galaxies lie close to Caldwell 72 should not be avoided when in the region. Just move 2½° to the north-northeast, and you should be able to sweep up tenth magnitude NGC 1808 and tenth magnitude NGC 1792. The former lies half a degree northeast of the latter. NGC 1808 has dimensions of 6.3 by 4.3 arcmin, and NGC 1792 is smaller at 5.6 × 2.9 arcmin. On November 15, 1993, Wischnjewsky discovered a seventeenth magnitude type Ia supernova in NGC 1808.

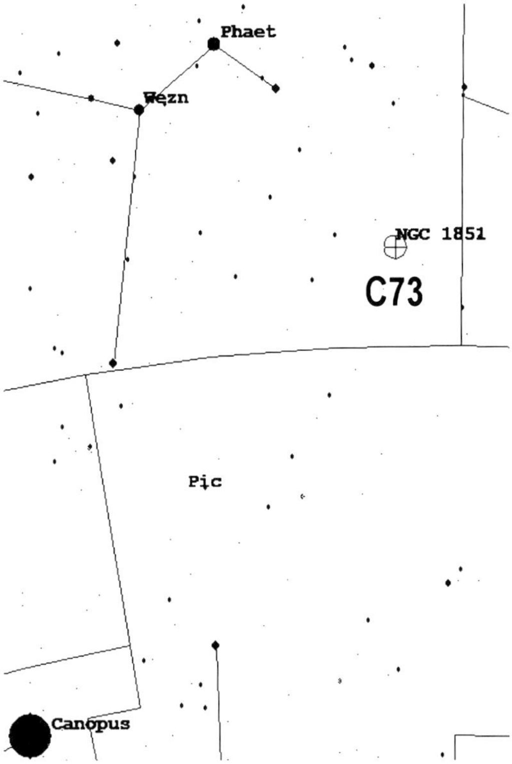

The chart shows the positions of Caldwell 73 in Columba in relation to the stars Phaet, Wezn and the dazzling Canopus.

NGC 1851 from the Digitized Sky Survey. The field is 15 arcmin wide.

Caldwell 74

NGC 3132
Southern Ring Nebula in Vela
Magnitude: 9.4
Size: 85″ × 55″
Mag/Sec: 19
RA: 10 h 07 min 02 s
Dec: –40° 26′
Circumpolar: below 50°S
Starhop: 7°E from ψ Velorum
Best Months: February to May, S. Hemisphere
Distance: 2,000 light-years
Physical Size: ~10 light-months
Age: ~10,000 years?
Best Visual Aperture: 15 cm and larger with high powers
Best Visual Filter: OIII or UHC filter
CCD/DSLR Advice: Try LRGB with Ha (L) and OIII (G)
Celestial Neighbors: Faint galaxies to E & SE
Miscellaneous: Lyra's ring, 73° further south!

Although the planetary nebula NGC 3132 is barely an eighth of the linear dimensions of the magnificent Helix Nebula (Caldwell 63), two magnitudes fainter, and almost four times more distant, it is still an excellent example of this category of deep sky object. It is also very worthy of its nickname of the "Southern Ring Nebula." The northern counterpart is, quite obviously, M57 in Lyra, which has virtually identical dimensions but is slightly brighter. In theory, an observer living near the equator could examine M57 and NGC 3132 on the same night in, say, May, with the southern Vela object low in the southwest and M57 rising in the northeast. There is a 73° declination difference and an almost 9-h right ascension difference, but it is possible. It is not known if any equatorial deep sky observers have achieved this feat. With two identical Dobsonians looking in opposite directions, you could walk from one eyepiece to the next comparing the two in short succession.

One thing you will notice if you move to the equator and try this feat is that the central star of the Caldwell object looks much brighter than that of M57. In fact, this is an illusion. The obvious tenth magnitude star in the center of NGC 3132 is not the star responsible for making the gases in the nebula fluoresce; it is actually that same star's sixteenth magnitude binary companion that does the job, and despite being fainter visually it has a temperature of 100,000 K. The stunning picture of the Southern Ring Nebula taken by the Hubble Space Telescope (opposite) shows both stars separated by less than 2 arcsec (roughly a light-week at the nebula's distance) as well as a vertical tendril splitting the blue inner cavity into two parts. Also obvious in that image is the nebula's multiple overlapping flat tire appearance, probably caused by the eccentric rotation of the stars about each other. This multiple ring appearance has given the nebula another nickname, namely "The Eight-Burst Nebula." Through a telescope you will see much less, of course, namely, a ghostly ellipse with a central star that is the wrong star!

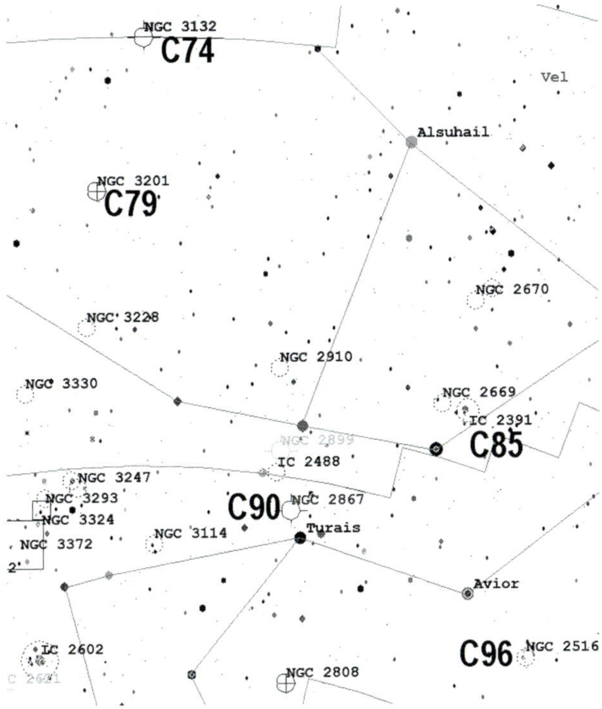

NGC 3132 (C74) lies very close to the Vela/Antlia border at the top of the chart. Caldwell's 79, 85, 90, and 96 are also shown.

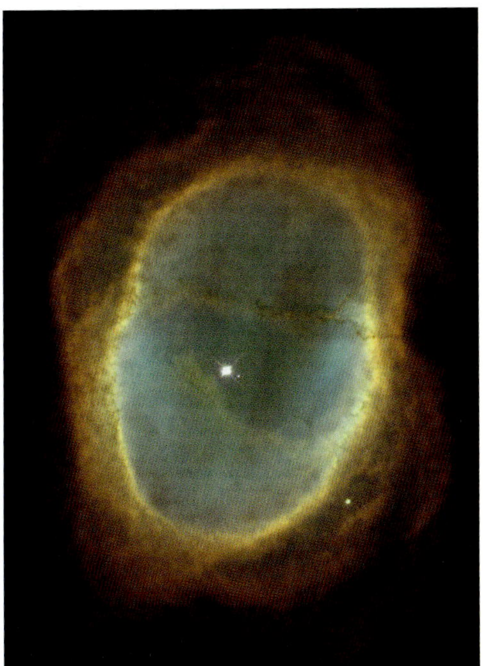

A high resolution Hubble Space Telescope view of the planetary nebula NGC 3132.

Caldwell 75

NGC 6124
Open Cluster in Scorpius
Magnitude: 5.8
Size: 40′
Mag/Sec: Not relevant
RA: 16 h 25 min 18 s
Dec: –40° 40′
Circumpolar: below 50°S
Starhop: 6½°S and 5°W from ε Scorpius
Best Months: May to August, S. Hemisphere.
Distance: 1,500 light-years
Physical Size: ~17 light-years
Age: ~150 million years
Best Visual Aperture: Big binoculars work well
Best Visual Filter: No filter needed
CCD/DSLR Advice: Easy target
Celestial Neighbors: NGC 6153 (planetary); NGC 6139 (glob.)
Miscellaneous: Just a dozen ninth magnitude stars?

Tucked away in the southwestern corner of Scorpio, near the borders with Lupus (to the West) and Norma (to the south), you may stumble across the open cluster NGC 6124, roughly 6½° south and 5° west of the Scorpion's magnitude 2.3 backbone star, ε Scorpius. At first glance the fifth magnitude rating of this cluster looks most interesting, but statistics can be misleading. Note the size: almost three-quarters of a degree. The cluster has the same integrated magnitude as Caldwell 71 (remember – is it an open cluster or a globular?) but spread over four times the sky and made up not of a hundred thirteenth or fourteenth magnitude stars but mainly a dozen ninth magnitude stars, which tend to prevent the fainter stars catching your eye. Add this to the fact that the cluster lies within the Milky Way's western boundary, and the borders of NGC 6124 become impossible to define. It looks just like a dozen ninth magnitude stars that have briefly joined together for a conference. The human eye/brain combination is very good at imagining patterns exist in clusters such as NGC 6124, and patterns containing arms and legs are the most likely to be visualized. Outside the cluster's official (40 arcmin diameter) eastern boundary you can imagine another cluster, and half a dozen ninth to eleventh magnitude stars reside there, too. Thus, depending on aperture, you may see one cluster, an elliptical cluster, or more than one cluster, but the feature that convinces you that this object deserves a Caldwell number is that main loose grouping of a dozen ninth magnitude stars. Other celestial attractions live in the immediate vicinity, too. Slightly more than a degree to the east-northeast you will find the twelfth magnitude planetary nebula NGC 6153, and 2° to the north (and slightly east) you will also find the globular cluster NGC 6139.

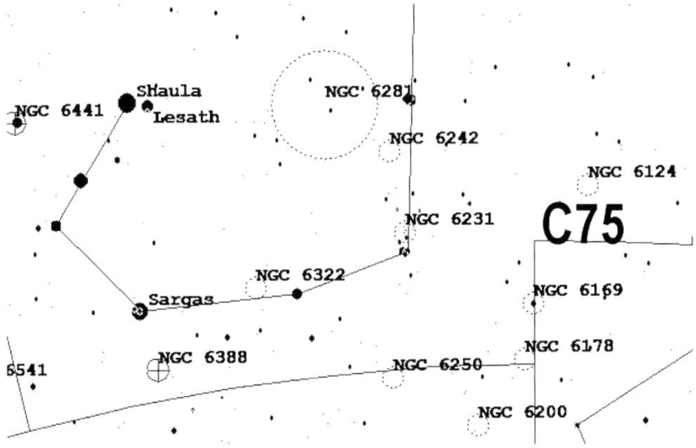

The chart shows the location of Caldwell 75 in relation to the Scorpion's tail, to the East.

NGC 6124 from the Digitized Sky Survey. The field is 15 arcmin wide.

Caldwell 76

NGC 6231
False Comet Cluster in Scorpius
Magnitude: 2.6
Size: 14′
Mag/Sec: Not relevant
RA: 16 h 54 min 12 s
Dec: –41° 50′
Circumpolar: below 49°S
Starhop: 7½°S and ¾°E from ε Scorpius
Best Months: May to August
Distance: 6,000 light-years
Physical Size: ~25 light-years
Age: ~7 million years
Best Visual Aperture: Big binoculars work well
Best Visual Filter: No filter needed
CCD/DSLR Advice: Easy target
Celestial Neighbors: Trumpler 24, C69 ~5°NE
Miscellaneous: The fake Halley of 1986!
Chart: See Caldwell 78

Not far to the east of the previous Caldwell object, we find another open cluster of remarkably different appearance. This cluster can also be reached from ε Scorpius (move 7½° south and three-quarters of a degree east), but it is easier to just look for the start of the eastward kink of the Scorpion's tail, find the double star ζ Scorpio, and travel just more than half a degree north. The whole region is quite eye-catching even to the naked eye as, in addition to NGC 6231, to the north-northeast the young cluster Trumpler 24 (part of the Sco OB1 Association) trails away, looking almost like a naked eye comet with NGC 6231 as the head.

Back in April 1986 this author traveled south to Tenerife to get a view of comet Halley at its peak. It was my first astronomical holiday. At the start of the month the comet was third magnitude and passed close to the base of the Scorpion. On my first glance at the region, and my first ever view of the whole of Scorpius, I could not tell which fuzzy object with a tail was the comet! It took a pair of binoculars to sort me out. The "head" of my "fake Halley" was, of course, NGC 2361, allotted the number 76 in Patrick's catalog.

Caldwell 76 is an impressive box of gems in a small telescope, looking a bit like luminous snooker or billiard balls in their triangular rack just before a game starts. The integrated magnitude of the cluster is an impressive 2.6 because this triangle (nose pointing southwest) contains more than a dozen stars of fifth to seventh magnitude and a dozen more of eighth, ninth, or tenth magnitude. Scores more stars between eleventh and thirteenth magnitude fill in the background glow surrounding the triangle out to roughly 7 arcmin from the cluster center – a very pretty object in any aperture.

This fine image of NGC 6231 was taken by Josef Pöpsel and Stefan Binnewies on 2006 May 4 from the Capella Observatory, then located at Amani Lodge, Kupferberg near Windhoek, Namibia. A Ganymed 60 cm-Hypergraph was used at primary focus, *f*/3. SBIG STL-11000M CCD. RGB 4 × 300 s: 3 × 300 s: 3 × 300 s. Image by kind permission of Stefan Binnewies and Josef Pöpsel.

Caldwell 77

NGC 5128
Galaxy Centaurus A
Magnitude: 6.7
Size: 17′ × 13′
Mag/Sec: 21
RA: 13 h 25 min 30 s
Dec: −43° 01′
Circumpolar: below 47°S
Starhop: 6¼°S and 1°E from ι Centauri
Best Months: March to June
Distance: 12 million light-years
Physical Size: ~60,000 light-years
Age: Billions of years
Best Visual Aperture: 10 cm and larger
Best Visual Filter: Deep sky filter may help
CCD/DSLR Advice: Awesome in mono or color
Celestial Neighbors: Omega Centauri is 4½°S
Miscellaneous: There's something a bit scarey about Centaurus A!

Surely every amateur astronomer must be familiar with the dramatic appearance of the extraordinary galaxy Centaurus A? No other galaxy in the sky looks quite like it. It lives just to the east of the center of the constellation Centaurus, and locating ι Centauri and then moving roughly 6° south and a degree east will bag it. An alternative strategy might be to bag the magnificent and unmissable globular cluster Omega Centauri first (see Caldwell 80) and move just 4½° north.

Centaurus A is a celestial powerhouse, spewing out radiation across the electromagnetic spectrum. The energy coming out of it at infrared and radio wavelengths is massive, and the same applies at the other end of the spectrum, namely in X-rays and gamma-rays. But, above all, it looks to be an absurdly violent place. Many edge-on galaxies have a dramatic dust lane in the center; one only has to think of M104, the Sombrero Galaxy in Virgo, NGC 891 (Caldwell 23) in Andromeda, or even the violent center of M82 in Ursa Major. However, Centaurus A looks far more violent than any of these. The sheer thickness of the dust lane stretching across the galaxy like an equatorial belt is extraordinary, but in Hubble images the fine central detail is so dense that the starfields and dust lanes look like an angry orange thundercloud about to explode.

Some amateurs have nicknamed Centaurus A the "Hamburger Galaxy"! Professional astronomers think the bizarre appearance of this celestial showpiece is a result of a collision with a smaller spiral galaxy. Despite not being the largest (in terms of its diameter) galaxy in the Caldwell catalog, Centaurus A is absolutely colossal in terms of mass. There may be a trillion solar masses in there (five times as much mass as in the Milky Way), and the supermassive black hole at the center may contain 200 million to a billion solar masses! On May 3, 1986, the prolific amateur supernova discoverer Robert Evans discovered a 12th magnitude Type Ia supernova in Centaurus A, the only one discovered in this bizarre galaxy.

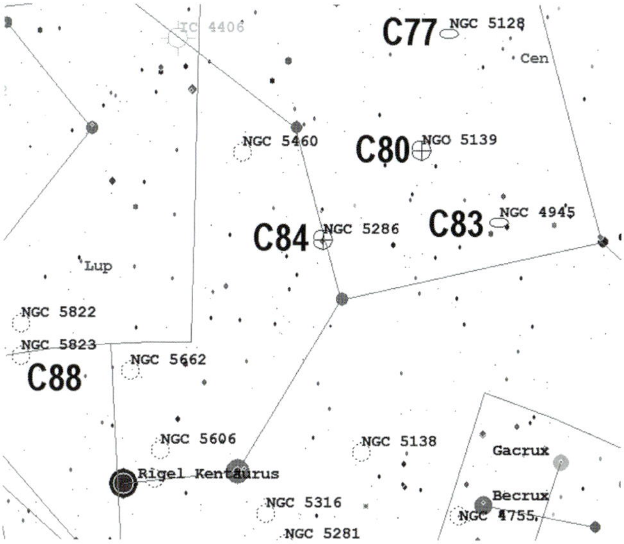

Centaurus A/NGC 5128, also known as Caldwell 77, is obvious in the *top right* of the chart. Part of the Southern Cross (*lower right*) and alpha plus beta Centauri (*lower center*) are also shown along with Caldwell's 80, 83, 84, and 88.

A splendid image of NGC 5128 taken by Daniel Verschatse at his Observatorio Antilhue, San Esteban, Chile. RCOS 14.5-in. (37 cm) Ritchey-Chrétien @ f/6.56 (Astrophysics 0.67× reducer). Astro-Physics AP1200-GTO mount. SBIG ST-10XE self-guided CCD. LRGB 120:90:90:90 mins (RGB binned 2 × 2) @ −15°C. 2004 March 20 @ 08:00 UTC. © Daniel Verschatse – Observatorio Antilhue – Chile.

Caldwell 78

NGC 6541
Globular Cluster in Corona Australis
Magnitude: 6.3
Size: 15′
Mag/Sec: 21
RA: 18 h 08 min 02 s
Dec: –43° 43′
Circumpolar: below 47°S
Starhop: ¾°S and 5½°E from Sargas (θ Sco)
Best Months: May to August, S. Hemisphere.
Distance: 22,000 light-years
Physical Size: ~100 light-years
Age: ~12 billion years
Best Visual Aperture: 10 cm and larger
Best Visual Filter: No filter advantage
CCD/DSLR Advice: Easy imaging target
Celestial Neighbors: Globular NGC 6496 lies 100′ WSW
Miscellaneous: A fine globular, somewhat neglected by imagers!

NGC 6541 is the first of four notable and even awesome consecutive Caldwell globulars in a row. We are back to the region of the sky not far from the Scorpion's sting, but this time we are east of that feature. Locate the bright (magnitude 1.9) star Sargas in the Scorpion's tail and slide 5½° east and almost a degree south. This will get you to the globular we seek, but some caution is required, as a fainter and smaller globular cluster (NGC 6496) right on the Scorpius/Corona Australis border is not far away. NGC 6541 is, however, a much more impressive offering. It was not all that many years ago that old globular clusters such as NGC 6541 were believed to be 16 or 17 billion years old. Any newcomer to astronomy may be rather confused by this, as most astronomers now agree that the universe is roughly 13.7 billion years old. Much attention has been focused on this dilemma in recent years, and there is a general belief that the oldest globulars are not now older than the universe. (Phew!) By placing the globular clusters at a greater distance than previously thought this makes their stars brighter and younger by enough of a margin to fit their ages into a 13.7 billion year old universe. NGC 6541 is relatively close, by globular standards, 22,000 light-years away, which makes all the difference from a visual perspective. As soon as individual stars can be resolved and held steady a globular will sparkle and take on a three-dimensional appearance. This fine globular cluster contains half a dozen 10th or 11th magnitude stars and a dozen more of twelfth magnitude. In a 30 cm or larger telescope these will positively sparkle. The globular becomes gradually denser as you move in towards the center, which contains a very tight, 2 arcmin in diameter, core – all in all, a splendid object.

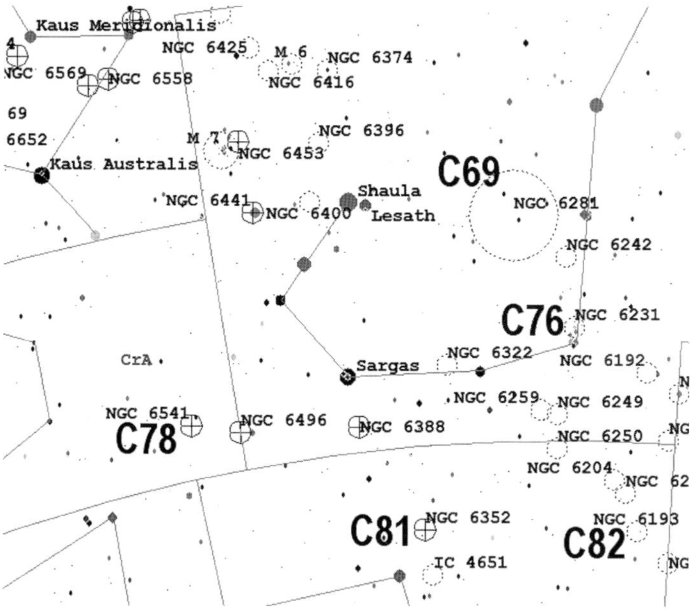

The chart shows the location of the bug nebula, Caldwell 69, in the Scorpion's tail, along with Caldwell objects 76, 78, 81, and 82 in Scorpio, Ara and Corona Australis.

NGC 6541 from the Digitized Sky Survey. The field is 15 arcmin wide.

Caldwell 79

NGC 3201
Globular Cluster in Vela
Magnitude: 6.7
Size: 20′
Mag/Sec: 22
RA: 10 h 17 min 37 s
Dec: –46° 25′
Circumpolar: below 44°S
Starhop: 12½°E & 3°S of λ Vel (Alsuhail)
Best Months: February to May, S. Hemisphere.
Distance: 17,000 light-years
Physical Size: ~100 light-years
Age: ~11 billion years
Best Visual Aperture: 10 cm and larger
Best Visual Filter: No filter advantage
CCD/DSLR Advice: Easy, but rewarding CCD target
Celestial Neighbors: Faint galaxies 2°NE
Miscellaneous: Silvery blue in color images, peppered with red giants
Chart: See Caldwell 74

Like Caldwell 78, Caldwell 79 is another relatively nearby sixth magnitude globular cluster with a physical diameter of 100 light-years. However, it lives in a different region of the sky and, although slightly larger in angular extent, is less dramatic than the previous object. NGC 3201 lies 12½° east and 3° south of the magnitude 2.2 star Alsuhail (λ Vel) or 3° north and 5° west of the magnitude 2.7 star μVelorum, the most easterly bright star in that constellation.

Trying to observe this globular on the same night as the "Southern Ring" Nebula (C 74) would be a sensible plan, as both objects are in Vela, just 6° apart. Although the globular contains a small number of stars as bright as twelfth magnitude (and one of eleventh), the bulk of its brightest stars are of 14th or 15th magnitude. Nevertheless, this is just sufficient to make any globular sparkle in very big apertures. Caldwell 79 has a catalog magnitude of 6.7 and so is just below naked-eye visibility even from a dark site, unless you have superhuman vision. Coincidentally, a magnitude 6.5 star (GSC 8183 2965) sits just 25 arcmin south of the cluster. In a small pair of binoculars it may be an interesting exercise to defocus the star, until it appears 20 arcmin across, and see whether you think it looks fractionally brighter than the globular.

NGC 3201 was observed by Sir John Herschel at the Cape of Good Hope using an 18-in. *f*/13 telescope with a speculum metal mirror. His comments were recorded as "globular cluster, irregularly round, gbM (gradually brighter in the middle), not vm (very much) compressed, 6′, resolved into stars 13...15th magnitude." In CCD images C79 is a silvery blue color with countless numbers of those thirteenth to fifteenth magnitude stars easily resolved as steely blue suns.

New Zealand observer Geof Wingham obtained this superb image of the globular cluster NGC 3201 using a Celestron 9.25-in. (235 mm) at *f*/8 on a Losmandy G-11 Gemini mount and a one-shot color SBIG ST-2000XCM CCD camera.

Caldwell 80

NGC 5139
Omega Centauri Globular Cluster
Magnitude: 3.7
Size: 55′
Mag/Sec: 21
RA: 13 h 26 min 46 s
Dec: −47° 29′
Circumpolar: below 43°S
Starhop: ¼°S and 5°W from ζ Centauri
Best Months: March to June
Distance: 17,300 light-years
Physical Size: ~280 light-years
Age: ~12 billion years
Best Visual Aperture: Awesome in any instrument
Best Visual Filter: No filter advantage
CCD/DSLR Advice: Mindblowing in mono or color
Celestial Neighbors: Centaurus A lies 4½°N
Miscellaneous: Noted since Ptolemy's era
Chart: See Caldwell 77

Omega Centauri is generally regarded as the finest globular cluster in the night sky and it is a really spectacular sight, especially in a telescope. Undoubtedly, the other truly awesome globular, 47 Tucanae (see Caldwell 106), gives it some stiff competition and is virtually the same brightness, but most observers would just give Omega Centauri the vote. The globular sits 5° to the west of ζ Centauri, but anyone who can recognize the Centaurus shape can find it. At magnitude 3.7 it is a naked-eye object. Viewing this spectacular globular together with the equally dramatic galaxy Centaurus A (dealt with three objects ago, 4½° to the north) is virtually mandatory. Who would observe one and not the other, except a madman! Omega Centauri is awesome both because it is absolutely massive in real terms and because it is relatively nearby as globulars go. This combination makes it span almost one degree of the sky. Admittedly if you brought the globular NGC 2419 (Caldwell 25) in from its absurd 280,000 light-year distance it would look even bigger, but that is not an option. However, Omega Centauri is not just 280 light-years across; it is thought to contain several million stars, comparable to some of the smallest dwarf galaxies. One significant advantage that Omega Centauri has over 47 Tucanae is its declination. Sadly, many northern hemisphere observers cannot see either of these marvels, but, below 42° north, the former can theoretically be glimpsed, even if it is just a smidgen above the horizon. You have to be below 18° north to get a sniff of 47 Tuc, so it is invisible even from southern Florida or the Mediterranean. With myriads of 14th, 13th, and even 12th magnitude stars the view through any decent aperture is mesmerizing.

Another stunning image from Robert Gendler, again imaging remotely using a 14.5-in. (37 cm) *f*/9 RCOS Ritchey-Chrétien with field flattener, based in Western Australia. Paramount ME mount and SBIG STL-11000XM CCD. This is a two frame mosaic made up from four hours of imaging data. © Robert Gendler.

Caldwell 81

NGC 6352
Globular Cluster in Ara
Magnitude: 8.0
Size: 7′
Mag/Sec: 21
RA: 17 h 25 min 29 s
Dec: –48° 25′
Circumpolar: below 42°S
Starhop: 1½°N and 1°W from α Ara
Best Months: May to August
Distance: 19,000 light-years
Physical Size: ~40 light-years
Age: ~11 billion years
Best Visual Aperture: 15 cm and larger
Best Visual Filter: No filter advantage
CCD/DSLR Advice: Easy target
Celestial Neighbors: C82 is 7° to the W
Miscellaneous: Missed (!) by John Herschel
Chart: See Caldwell 78

Our fourth consecutive Caldwell globular lives in the constellation of Ara, the Altar and, once again, we are not far from the Scorpion's tail and sting as Ara's northern border is also Scorpio's southern border. After the magnitude 3.7 searchlight of Omega Centauri an 8th magnitude offering may seem rather puny, and it is! NGC 6352 is not small and faint because it is far away; it is actually about the same distance as Omega Cen. Itself. It is simply a puny globular, only 40 light-years in diameter.

Although you could use the Scorpion's sting star Sargas to hop to NGC 6352, the magnitude 2.8 star α Ara is much nearer; from there it is just a degree and a half north and a degree west to the globular. There is an open cluster in the region, too; IC 4651 lies just a degree due west of Ara and is significantly brighter. A beginner could confuse the two. Although NGC 6352 might be rather disappointing its relative proximity does mean that, as with Omega Centauri, many of its brighter stars can be resolved. These shine at 12th and 13th magnitude and compensate for the otherwise less-than-spectacular appearance of the object. With 40-cm apertures and larger instruments significant numbers of 15th magnitude stars will make the globular more resemble its photographic appearance. In CCD images, the backdrop of the Milky Way passing through Ara is very conspicuous and makes determining the edge of the globular a far from exact science. If you find Caldwell 81 rather dull take a trip 90 arcmin south to the other object mentioned earlier. The open cluster IC 4651 is actually larger in angular extent than the globular and spans 10 arcmin. It contains about 80 stars of magnitude 13 and brighter with a handful of tenth magnitude stars and one-ninth magnitude offering in the extreme northeast of its diameter. Some observers may find it the more attractive object at magnitude 6.9. However, a brighter open cluster in Ara is the next Caldwell offering.

NGC 6352 from the Digitized Sky Survey. The field is 10 arcmin wide.

Caldwell 82

NGC 6193
Open Cluster in Ara
Magnitude: 5.0
Size: 15′
Mag/Sec: Not relevant
RA: 16 h 41 min 18 s
Dec: –48° 46′
Circumpolar: below 42°S
Starhop: 1°N and 8¼°W from α Ara
Best Months: May to August
Distance: 4,300 light-years
Physical Size: ~20 light-years
Age: ~6 million years
Best Visual Aperture: Big binoculars and larger instruments
Best Visual Filter: No filter required
CCD/DSLR Advice: Easy target – deep images reveal nebulosity
Celestial Neighbors: NGC 6200, 6204, 6167, 6134 nearby
Miscellaneous: More here than just an open cluster!
Chart: See Caldwell 78

In the northwestern region of Ara, close to its western border with Norma, you will find the open cluster NGC 6193. At least, you will find it with a bit of care, because there are four other fainter open clusters within 2°, one of 6th magnitude, two of 7th magnitude, and one of 8th magnitude. But as our target is magnitude 5.0 it should stand out. Traveling a degree north and roughly 8° west from α Ara will get you to Caldwell 82. The nearest naked eye star is magnitude 4.5 ε Normae; from that star just hop 2½° east and 1¼° south. Caldwell 82 was listed as an open cluster in Patrick's catalog, first published in the December 1995 edition of *Sky & Telescope*. However, in any wide field image of the region it is obvious that there is far more in this environment than just one open cluster. The cluster forms the core of the vast Ara OB1 association of hot, young stars, which occupies a square degree of the sky.

NGC 6193 sits within a region of dense gas clouds and obscuring lanes of dust. Its brightest component is an O type star easily visible in the cluster center, shining at magnitude 5.6. It has a catalog name of HD 150136 and effectively illuminates the emission nebula NGC 6188. HD 150136 itself is a multiple system with a magnitude 6.8 companion only 8 arcsec to the west. A magnitude 7 star sits at the cluster's western edge, and a magnitude 8 star is just to the northeast of the cluster center. A dozen other 9th and 10th magnitude stars lie within the cluster, mainly to the east. The cluster itself is a pretty sight through a small telescope, but it needs a deep CCD image to truly reveal the nebular splendor of the region.

NGC 6193 from the Digitized Sky Survey. The field is 15 arcmin wide.

Caldwell 83

NGC 4945
The Tweezers Galaxy in Centaurus
Magnitude: 8.8
Size: 19′ × 3.5′
Mag/Sec: 22
RA: 13 h 05 min 24 s
Dec: –49° 28′
Circumpolar: below 41°S
Starhop: ½°S and 4°E from γ Centauri
Best months: March to June
Distance: 17 million light-years
Best Months: ~95,000 light-years
Age: Billions of years
Best Visual Aperture: 20 cm and larger
Best Visual Filter: Deep sky filter may help
CCD/DSLR Advice: Superb in mono or LRGB
Celestial Neighbors: Omega Centauri 4° to the NE
Miscellaneous: Worth patrolling for supernovae, despite edge-on
Chart: See Caldwell 77

For anyone unfamiliar with the southern skies prior to reading this book, it must be obvious by now that Centaurus is a superb constellation for the deep sky observer. The trillion solar mass radio galaxy Centaurus A and the magnificent globular cluster Omega Centauri are remarkable, but there is yet more to enjoy here, too, such as the splendid edge-on galaxy NGC 4945. The nearest bright star to this galaxy is γ Centauri. Centering that star and then moving ½ a degree south and 4° east will get you there. Alternatively, locating Omega Centauri and moving 4° southwest will get you there, too. At 19 arcmin in length this is yet another very long, thin spike in the telescope eyepiece. Two amateur astronomers from Brazil, C. Jacques and E. Pimentel, discovered a supernova in NGC 4945 on exposures made on Feb 8th and 10th in 2005 with a 30 cm Schmidt-Cassegrain. The telescope was sited at the CEAMIG/REA observatory and was part of BRASS, the Brazilian Supernova Search. The supernova reached a highly respectable magnitude 12.5 and was 407 arcsec west and 351 arcsec south of the galaxy center, placing it just above the lower right hand edge of the galaxy in "north-up" images. Through any telescope of 100-mm aperture or larger NGC 4945 is a rewarding splinter of light. In large apertures the tweezer-like appearance is obvious and caused by the obscuring central dust lane. In deep LRGB CCD images a wonderful gold core contrasts with the gray/blue disk and browny dust lane. A much fainter galaxy of magnitude 11.1, NGC 4976, is worth checking out if you have a large aperture telescope and are in this region. The dimmer fuzz lies only half a degree east of NGC 4945 and has an oval appearance (5.1′ × 2.6′).

This stunning image of NGC 4945 was taken by Josef Pöpsel and Beate Behle on 2005 May 10 from the Capella Observatory, then located at Amani Lodge, Kupferberg near Windhoek, Namibia. A Ganymed 60 cm-Hypergraph was used at primary focus, *f*/3, $f = 180$ cm. SBIG ST10-XME CCD. L(R + Hα)GB. L: 12×300 s 1×1 bin, R, Hα, G, B: 4×300 s. each, 2×2 bin. On the original image the fading supernova 2005af, three months past maximum, can be seen as a faint reddish star near the lower right tip of the galaxy; Image by kind permission of Stefan Binnewies and Josef Pöpse.

Caldwell 84

NGC 5286
Globular Cluster in Centaurus
Magnitude: 7.3
Size: 11′
Mag/Sec: 21
RA: 13 h 46 min 27 s
Dec: –51° 22′
Circumpolar: below 39°S
Starhop: 2°N and 1°E from ε Centauri
Best Months: March to June, S. Hemisphere.
Distance: 35,000 light-years
Physical Size: ~110 light-years
Age: ~12 billion years
Best Visual Aperture: 10 cm and larger
Best Visual Filter: No filter advantage
CCD/DSLR Advice: Easy target
Celestial Neighbors: Omega Centauri lies 5° to the NW
Miscellaneous: Rather neglected due to other Centaurus wonders
Chart: See Caldwell 77

The deep sky offerings from Centaurus just keep on coming! Our next object is yet another globular cluster; sadly, not in the same league as Centaurus A, but still worth a look. Draw an imaginary line between magnitude 2.3 ε Centauri and magnitude 2.5 ζ Centauri and slide 35% of the way from ε to ζ and you should spot Caldwell 84. Its location is enhanced by the presence of the magnitude 4.6 foreground star GSC 8274 2554 (M Centauri), which sits, by chance, at the extreme southeastern edge of the globular, where its stellar population fades into the sky background.

Although the bright star may be an excellent marker, in a telescope eyepiece it tends to dazzle any visual inspection of the cluster, whose core lies less than 5 arcmin to the northwest. M Centauri lies at roughly 260 light-years from Earth or 130× closer than the globular cluster. It has a luminosity 70 times as great as our own Sun and outshines the globular, from our vantage point, by a factor of 10. Of course, being 130× closer than if it were a cluster star provides a 130 squared advantage even without its very healthy luminosity. M Centauri is a binary star with the fainter component thought to orbit the primary in 437 days. The secondary star is only detectable spectroscopically. As for the globular itself, well, most of its stars are only of 15th magnitude or fainter, but there are enough 13th and 14th magnitude stars to allow NGC 5286 to sparkle a bit even in modest apertures. In some respects the globular resembles M68 in the constellation of Hydra, which has a similar magnitude and diameter and, at a declination of –26° 45′, will be more accessible to northern hemisphere observers.

A splendid image of NGC 5286 taken by Daniel Verschatse at his Observatorio Antilhue, San Esteban, Chile. RCOS 14.5-in. (37 cm) Ritchey-Chrétien @ *f*/9 (prime focus). Astro-Physics AP1200-GTO mount. SBIG ST-11000 self-guided CCD. LRGB 50:12:12:12 min @ –20°C. 2005 April 11 @ 06:00 UTC. © Daniel Verschatse - Observatorio Antilhue – Chile.

Caldwell 85

IC 2391
Omicron Velorum Cluster in Vela
Magnitude: 2.5
Size: 60′
Mag/Sec: Not relevant
RA: 08 h 40 min 18 s
Dec: –52° 55′
Circumpolar: below 38°S
Starhop: 1¾°N and ½°W of δ Velorum
Best Months: January to April
Distance: 500 light-years
Physical Size: ~9 light-years
Age: ~50 million years
Best Visual Aperture: Big binoculars and larger instruments
Best Visual Filter: No filter required
CCD/DSLR Advice: Easy, even with telephoto lenses
Celestial Neighbors: NGC 2669 < 1°E
Miscellaneous: Poor man's Pleiades and with no nebulosity, either
Chart: See Caldwell 74

Some 2 h of right ascension east of the brilliant star Canopus the constellation of Carina the Keel meets Vela the Sails. It will not have escaped the reader's notice that both these items are parts of a ship. In 1752, the French astronomer Nicolas Louis de Lacaille subdivided the giant constellation Argo Navis (after *Argo*, the ship used by the mythological Jason and the Argonauts) into Carina the Keel, Puppis the Poop, and Vela the Sails. If Argo Navis still existed it would be the largest constellation in the sky, even larger than Hydra. Where the keel meets the sails, in the southwest corner of Vela, you may stumble across the bright Omicron Velorum Cluster. The nearest bright naked-eye star to the field is δ Velorum; from there just hop 1¾° north and ½° west. The Omicron Velorum Cluster is dominated by half a dozen bright stars and, in some ways, is a feeble version of the Pleiades, but without the distinctive pattern and electric blue nebulosity. At first glance in binoculars you may just see three distinctive stars forming a bent leg shape with a double star to the southeast and, maybe a 6th magnitude star to the southwest. The stars in the bent leg are Omicron (o) itself (at magnitude 3.6, the brightest), a magnitude 5.5 star to the south, and magnitude 5.2 NZ Velorum. The double star components are HY and KT Vel (magnitude 4.8 and 5.5). Apart from these stars and the magnitude 6 star already mentioned there are just four magnitude seven stars, a few 8th and 9th magnitude stars, and the rest are so much fainter that you wonder if they are cluster stars or part of the general background. With further study in big binoculars, or a small telescope, it is obvious that there are too many 10th and 11th magnitude stars for sheer chance, but, to be honest, the really bright stars dominate this cluster, which would be rather bland without them.

IC 2391 from the Digitized Sky Survey. The field is 15 arcmin wide.

Caldwell 86

NGC 6397
Globular Cluster in Ara
Magnitude: 5.7
Size: 25′
Mag/Sec: 21
RA: 17 h 40 min 41 s
Dec: –53° 40′
Circumpolar: below 37°S
Starhop: 3¾°S and 1½°E of α Ara
Best Months: May to August
Distance: 7,500 light-years
Physical Size: ~55 light-years
Age: ~11 billion years
Best Visual Aperture: 10 cm and larger
Best Visual Filter: No filter advantage
CCD/DSLR Advice: Easy with any imaging system
Celestial Neighbors: Globular NGC 6584 is 6°ENE
Miscellaneous: Superb, sparkling and nearby globular
Chart: See Caldwell 89

If you live in the northern hemisphere, you may not have heard of NGC 6397, and yet it is one of the best globular clusters in the sky. There is one reason for this: its distance. NGC 6937 is the third closest globular cluster to our Solar System. At an estimated 7,500 light-years, it is probably 300 light-years further away than the well known globular M4, but with the uncertainty in these types of measurements they are, essentially, at the same approximate distance. The closest globular cluster is, as of 2006, supposedly the faint example FSR 1767 (or 2MASS-GC04), which is estimated at only 4,900 light-years distant. Never heard of FSR 1767? That is because it is a tiny low mass globular in heavily crowded Scorpius, where interstellar absorption knocks 6.2 magnitudes off that object. To all intents and purposes M4 and NGC 6397 are the joint closest bright globulars. At magnitude 5.7 NGC 6397 is one of the brightest globulars, too, despite its rather small 55 light year diameter. The proximity of this globular cluster means that its brightest stars are easily visible in modest telescopes. Within its 25 arcmin diameter lie ten stars as bright as 9th or 10th magnitude, and more than a hundred others of 11th to 13th magnitude. This globular really will sparkle in the eyepiece, as its proximity and bright stars just tip it into the telescopic category of "something special." To find Caldwell 86 you need, once more, to visit the constellation of Ara the Altar and the star α Ara. From there it is a short hop 3¾° south and 1½° east. Alternatively it sits roughly 10½° south of our old friend Sargas in the Scorpion's tail. The Hubble Space Telescope has carried out a detailed study of NGC 6397 and identified many "blue straggler" stars as well as a millisecond orbital period neutron star, rotating 274 times a second and orbiting a bloated red star.

A stunning image of NGC 6397 taken by Daniel Verschatse at his Observatorio Antilhue, San Esteban, Chile. RCOS 14.5-in. (37 cm) Ritchey-Chrétien @ f9. Astro-Physics AP1200-GTO mount. SBIG STL-11000 self-guided CCD. LRGB 60:20:20:25 min (individual exposures 5 min) –25°C. 2005 June 5 @ 06:00 UTC. © Daniel Verschatse – Observatorio Antilhue – Chile.

Caldwell 87

NGC 1261
Globular Cluster in Horologium
Magnitude: 8.3
Size: 7′
Mag/Sec: 21
RA: 03 h 12 min 15 s
Dec: –55° 13′
Circumpolar: below 35°S
Starhop: 2°N and 13°E from Achernar (α Eridani)
Best Months: September to December
Distance: 55,000 light-years
Physical Size: ~110 light-years
Age: ~12 billion years
Best Visual Aperture: 20 cm and larger
Best Visual Filter: No filter advantage
CCD/DSLR Advice: Easy to image
Celestial Neighbors: Three v. faint galaxies on southern edge
Miscellaneous: Can be a challenge to track down

After the nearby NGC 6937, this next globular may be rather disappointing. NGC 1261 is 7× further away and 10× fainter. Physically it is a bigger object, but at 55,000 light-years it would need to be a monster to blow you away at the eyepiece. Not only is this eighth magnitude globular rather dismal, it may well be one of the toughest to track down unless you are using a "Go To" telescope or setting circles. Horologium (The Pendulum Clock) is a very uninspiring constellation, and starhopping from nearby bright markers to the faint globular is just not possible. There are no bright markers! The brightest star within 10° of the globular is magnitude 5.2 ζ Horologium. From there NGC 1261 is 4½° east and 40′ south. Alternatively, if you go to zero magnitude Achernar and set your setting circles to 1 h 37.7 min –57° 14′ and then push the telescope to 03 h 12.3 min –55° 13′ you should find the globular. A conspicuous 9th magnitude star sits 4 arcmin northeast of the cluster's core, just outside the faintest outer edges of the cluster. Another 9th magnitude star lies 13 arcmin west of the top (northern) edge of the cluster. Unfortunately, most of this 8th magnitude globular's stars are only 16th or 17th magnitude, making the resolution of individual stars virtually impossible except with the sort of monster Dobsonians seen at US star parties. But for imagers Caldwell 87 offers a bit more. Just off the globular's southern boundary lie three faint galaxies. To the southwest magnitude 15.3 PGC 11916 is overlapped by the smaller, but fractionally brighter (15.1) PGC 11918. To the southeast you may record magnitude 16.4 PGC 145348.

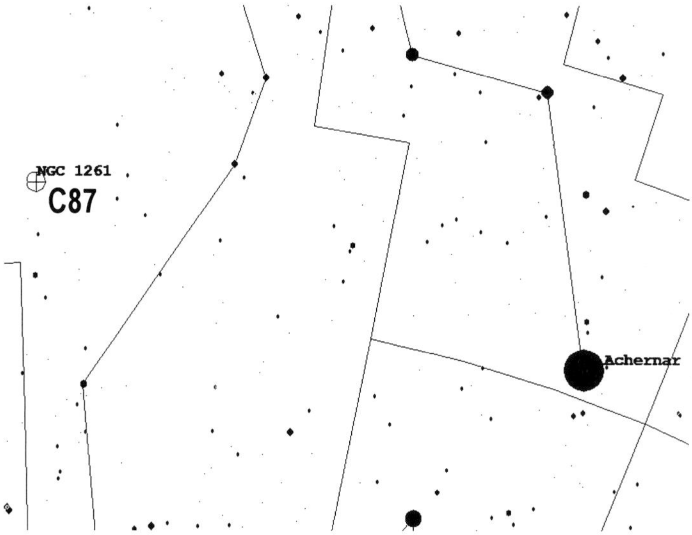

The position of the globular cluster Caldwell 87 in Horologium with respect to the bright star Achernar is shown.

NGC 1261; another superb image by Daniel Verschatse from his Observatorio Antilhue, San Esteban, Chile. RCOS 14.5-in. (37 cm) Ritchey-Chrétien @ f9. Astro-Physics AP1200-GTO mount. SBIG STL-11000 self-guided CCD. LRGB 50:20:20:20 min –20°C. 2006 August 21 @ 09:30 UTC. © Daniel Verschatse - Observatorio Antilhue – Chile.

Caldwell 88

NGC 5823
Open Cluster on Circinus/Lupus border
Magnitude: 7.9
Size: 12′
Mag/Sec: Not relevant
RA: 15 h 05 min 24 s
Dec: −55° 36′
Circumpolar: below 35°S
Starhop: 5¼°N and 3½°E from Rigel Kentaurus (α Centauri)
Best Months: April to July
Distance: 3,500 light-years
Physical Size: ~12 light-years
Age: ~800 million years
Best Visual Aperture: 15 cm and larger
Best Visual Filter: No filter required
CCD/DSLR Advice: Another easy target
Celestial Neighbors: Open cluster NGC 5822, 1°N
Miscellaneous: Constellation border straddler
Chart: See Caldwell 77

The open cluster listed as Caldwell 88 lies mainly in the constellation Circinus (the Compass), but its northern edge just makes it into Lupus (the wolf), just a degree below another open cluster, NGC 5822. Thus, there is some room for confusion when sweeping the area. Indeed, within 5° of our target there are six open clusters; as well as NGC 5822, NGC 5925 sits to the east, 5749 to the west, 5764 to the northwest, and the pair of 5622 and 5715 to the southwest. The region is probably best located by reference to the unmistakable zero magnitude duo of Alpha (α) and Beta (ß) Centauri, only 4½° apart in Centaurus. Alpha Centauri is, of course, the closest star *system* to Earth, lying only 4.4 light-years away. One fainter component of that system, Proxima, is 0.2 light-years closer. From Alpha (also known as Rigel Kentaurus) it is just 5¼° north and 3½° east to our target.

John Herschel observed this cluster in April 1836 and described it as a fine large cluster of separate stars of 13th to 14th magnitude, 10′ in diameter and not much compressed in the middle, "nearly fills the field." In fact, although rather faint stars do dominate the cluster a couple of 10th magnitude stars and a dozen of 11th are contained within NGC 5823's diameter. The cluster mentioned earlier, NGC 5822, just a degree further north, is a much better object of 6th magnitude and contains dozens of 10th magnitude stars and more than a hundred stars of 12th and 13th magnitude. It is also three times the diameter of the former cluster. Only a drug-crazed lunatic would observe NGC 5823 and not have a look at its neighbor and conclude the neighbor was the better object!

NGC 5823 from the Digitized Sky Survey. The field is 15 arcmin wide.

Caldwell 89

NGC 6087
Open Cluster S Normae
Magnitude: 5.4
Size: 15′
Mag/Sec: Not relevant
RA: 16 h 18 min 54 s
Dec: −57° 54′
Circumpolar: below 33°S
Starhop: 2°S and 5½°W from ζ Ara
Best Months: April to July
Distance: 3,300 light-years
Physical Size: ~14 light-years
Age: ~100 million years
Best Visual Aperture: Big binoculars and larger instruments
Best Visual Filter: No filter required
CCD/DSLR Advice: Dead easy target
Celestial Neighbors: C 95 (NGC 6025) lies 5°SW
Miscellaneous: Original Caldwell mis-identification with NGC 6067

It is but a modest journey east and slightly south to our next Caldwell object and yes, you have guessed it, it is another open cluster. We are now so far south in the celestial sphere that NGC 6087 never sets from southern hemisphere locations as far north as 33° south; thus, from South Africa and south Australia Caldwell 89 is a permanent night-time feature. The adjoining constellation of Ara, to the east, offers the nearest bright naked-eye stars. From magnitude 3.1 ζ Ara the starhop is 2° south and 5½° west to arrive at the S Norma cluster. In binoculars a distinctive, slightly curved line of stars leads you into the cluster from 2° out to the west. These are stars of magnitude 4.7 (westernmost), 5.6, and 5.6 again (a third of a degree west of the cluster).

Norma is awash with binocular magnitude open clusters. As well as eight within the boundaries a further three straddle the border regions with Ara and Triangulum; one of these, NGC 6025, we come to later, as Caldwell 95. In the original Caldwell catalog listing (on page 40 of the December 1995 *Sky & Telescope*) object 89 was listed as the "S Nor Cluster" but the NGC number was listed, erroneously, as 6067, not 6087. In fact, 6067 is one of those other Norma clusters, roughly 4° to the north. Bearing in mind the common name of this cluster, it will not surprise anyone that the dominant star within it is S Normae, shining at a magnitude of between 6.1 and 6.8. S Normae is a Cepheid variable star with a brightness that varies over a period of 9.8 days. Apart from this variable the brightest other star in the cluster is magnitude 7.9, 2 arcmin to the south, although there is a magnitude 7.5 star on the extreme western edge, which is the brightest star in a row leading south from that edge. A few 9th magnitude stars and two dozen of 12th magnitude make up the remaining space.

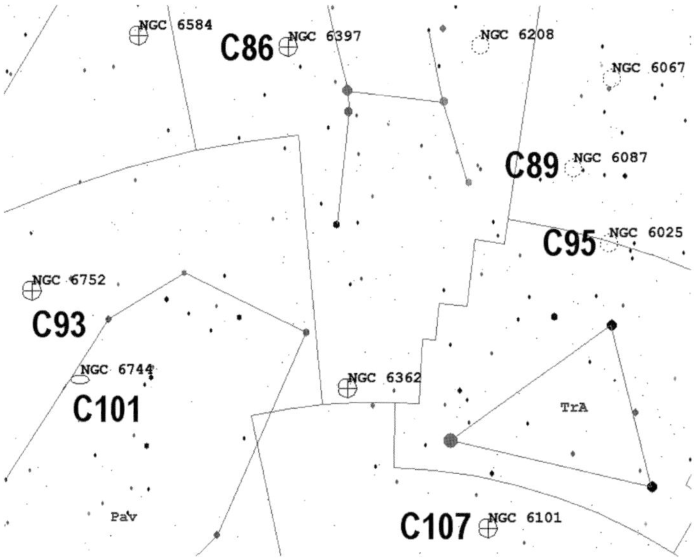

The chart shows a large swath of southern hemisphere sky including the constellations Pavo, Ara, Norma, Triangulum Australe and Apus. As well as C89, five other Caldwell objects are labelled.

NGC 6087 from the Digitized Sky Survey. The field is 15 arcmin wide.

Caldwell 90

NGC 2867
Planetary Nebula in Carina
Magnitude: 9.7
Size: 11″
Mag/Sec: 15
RA: 09 h 21 min 25 s
Dec: –58° 19′
Circumpolar: below 32°S
Starhop: 1°N and ½°E of Turais (ι Carinae)
Best Months: January to April
Distance: 5,500 light-years
Physical Size: ~15 light-weeks
Age: ~5,000 years
Best Visual Aperture: 25 cm and larger
Best Visual Filter: OIII or UHC
CCD/DSLR Advice: Tiny, stellar and difficult object
Celestial Neighbors: Open cluster IC 2488 lies 80′ to NNE
Miscellaneous: Best scrutinised by the Hubble Space Telescope!
Chart: See Caldwell 74

We have finally arrived at Carina the Keel, after dallying just above its northern border when we dropped in on Caldwell 85. The region we seek is dead easy to find, as the magnitude 2.2 star Turais is close by. Just find that star (ι Carinae), and the planetary nebula is a degree to the north and just half a degree to the east. Many planetary nebulae are small, but with a relatively high surface brightness, and NGC 2867 is no exception; in fact, it is *very*, *very* small. In fact, it is so small that even if it were a planet in our Solar System you would not expect to see many details on a disc of that size. Small telescopes will show NGC 2867 as a star at low and medium magnifications. At high powers, if you know that this is the planetary nebula in advance, you may sense the star is bloated. But really, this is an object best seen, and imaged, with huge telescopes and very high powers. The problem is, at 11 arcsec in diameter, atmospheric seeing, the bane of true planetary observing, will take its toll. There is another issue here too, namely, the structure within NGC 2867 is woolly and on a very small scale. Ironically, this planetary is bright and has a high surface brightness. Some have even reported a bluish color. So it is easy to observe, but almost impossible to see it for what it is.

To be honest, this is an object that only the Hubble Space Telescope can do justice, too, but even in images with that beast it looks a bit like a soggy biscuit with the first bite taken out of it. Caldwell 90 does have a central star, but, at 15th magnitude and immersed in the soggy biscuit nebulosity, no amateur will ever see it with any form of backyard telescope. Professional astronomers have determined that the nebula has a high nitrogen and carbon content.

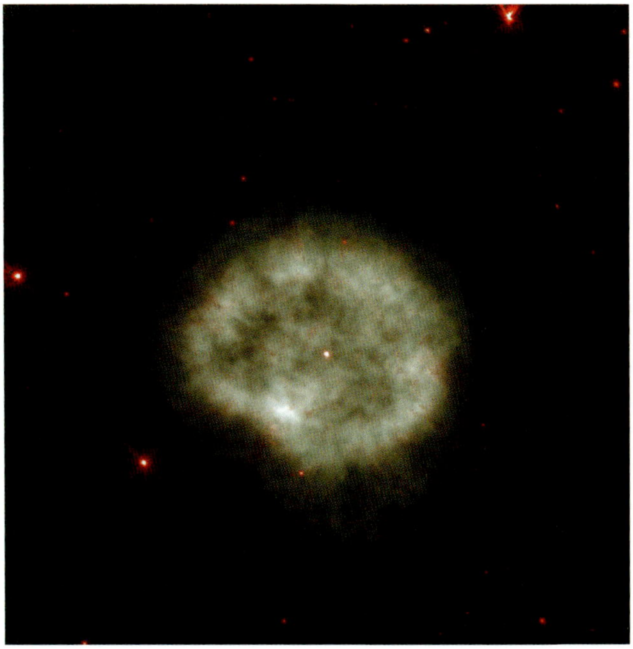

This staggering image of NGC 2867 was taken by the Hubble Space Telescope. The detail is incredible bearing in mind the object only spans eleven arc-seconds.

Caldwell 91

NGC 3532
The Pincushion Cluster in Carina
Magnitude: 3.0
Size: 50′
Mag/Sec: Not relevant
RA: 11 h 30 min 30 s
Dec: –58° 44′
Circumpolar: below 32°S
Starhop: 4¼°N and ½°W from λ Centauri
Best Months: March to June
Distance: 1,600 light-years
Physical Size: ~23 light-years
Age: ~300 million years
Best Visual Aperture: Big binoculars will suffice
Best Visual Filter: No filter required
CCD/DSLR Advice: Magnificent and easy with a 1° field
Celestial Neighbors: Eta Carinae neb (C92) 2°SW
Miscellaneous: A glittering marvel!

Roughly midway between our last Caldwell object and the famous "Southern Cross" lives one of the best star clusters in the sky: Carina's "Pincushion Cluster." The Pincushion is tucked away in the northeastern corner of Carina with the Vela border only a degree further to the northwest and the Centaurus border only a degree further to the northeast. The cluster is equidistant from the 3rd magnitude stars PP Carinae (to the southwest) and λ Centauri to the southeast. From the latter the offsets are 4¼° north and ½° west. The Pincushion is a glorious sight in any small telescope. Almost 700 stars reside in the cluster's 23 light-year diameter, and it is easily visible to the naked eye as a 50 arcmin fuzz. Just outside the cluster's southeastern boundary you will spot the Cepheid variable V382 Carinae, which shines at an average magnitude of 3.9 and is at least twice the distance of the cluster itself. Within the cluster the brightest stars are somewhat fainter but spectacular in number. One star of magnitude 6 lies to the northeast of the cluster center but there are dozens of seventh and eighth magnitude, concentrated mainly in a pincushion shape roughly half a degree in width (east-west) but maybe a quarter of a degree in height. The backdrop to this rectangle and, indeed, the whole 50 arcmin circle of the cluster is filled with hundreds more stars from 9th to 13th magnitude. There is nothing in the northern hemisphere that is equivalent to this magnificent cluster. While there are easy naked-eye clusters like the Pleiades, the Hyades and the Beehive they tend to be dominated by a handful of naked-eye stars. By comparison, NGC 3532 is crammed full of glittering stars that, while they are just below naked-eye visibility, hit you in the face as soon as some optical aid is applied. Some would say that the double cluster in Perseus comes nearest to approaching the Pincushion; its stars are much fainter, but the sense of awe when it drifts into the eyepiece field is very similar. Caldwell 91 is certainly one of the very best Caldwell choices.

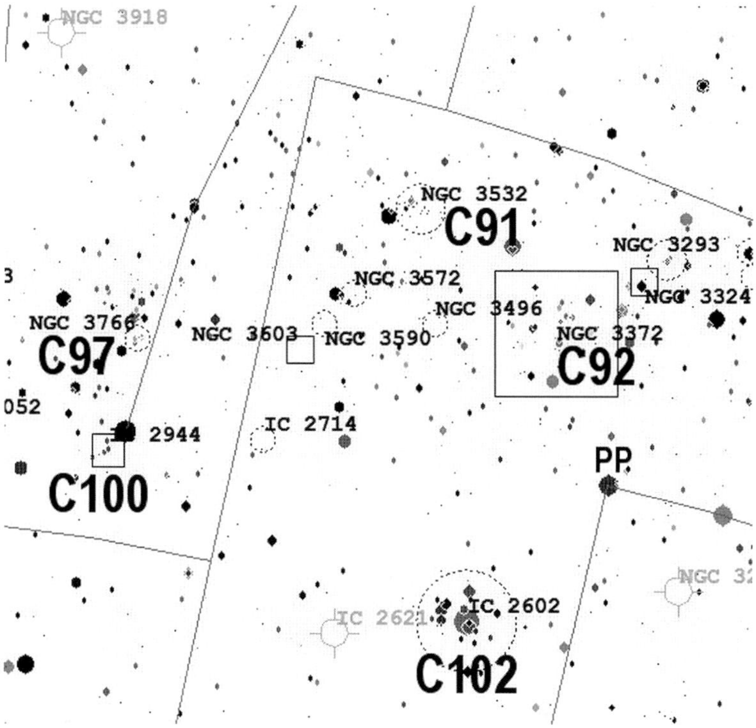

A total of five superb deep sky Caldwell objects, including Caldwell 91, are littered throughout the southern hemisphere constellations of Centaurus and Carina. This chart overlaps the one on the Caldwell 94 page.

NGC 3532 from the Digitized Sky Survey. The field is 30 arcmin wide.

Caldwell 92

NGC 3372
Eta Carinae Nebula
Magnitude: ~4.8
Size: 120′
Mag/Sec: 24
RA: 10 h 43 min 48 s
Dec: –59° 52′
Circumpolar: below 31°S
Starhop: 1¾°N and 1½°E from PP Carinae
Best Months: February to May
Distance: 7,500 light-years
Physical Size: ~260 light-years
Age: Eta Carinae < 3 million years old
Best Visual Aperture: Binoculars and telescopes of all sizes
Best Visual Filter: UHC for nebula and stars
CCD/DSLR Advice: Try ~500-mm lens + H-alpha filter
Celestial Neighbors: Pincushion (C91) 2°NE
Miscellaneous: A deep sky observer's wonderland – truly amazing!
Chart: See Caldwell 91

Spectacular Caldwell objects seem to be a bit like London buses. There are none for ages and then you get two in a row! Eta Carinae and its nebula are remarkable objects visually, cosmologically and historically. It is hard to know where to start when describing this object. Firstly, the nebula and star are right next door to the previous Caldwell object. Move 2½° west-southwest from the Pincushion and you are right in the heart of the Eta Carina nebula. Alternatively, hop 1¾° north and 1½° east from PP Carinae. Eta Carinae, the 4th to 7th magnitude star (its range in recent years) at the heart of this huge nebula is an enormous object with a mass of somewhere between 100 and 150 solar masses. There are probably no more than 30 or 40 stars this large in our entire Milky Way Galaxy (containing 200 billion stars). Eta Carinae is an unpredictable variable star that peaked in brightness in April 1843 at magnitude –0.8, making it second in brightness only to Sirius, despite being almost 1,000 times more distant! This level of outburst would normally be associated with such a star suffering a core-collapse supernova event. But Eta Carinae survived, although professional astronomers think it will go supernova in the near cosmological future – maybe in a million years, but maybe tomorrow. One of the Hubble Space Telescope's most awesome images is of the bipolar lobes from that 1843 explosion, the famous Homunculus Nebula, now spanning some 9 light-months after 165 years of expanding out from the massive star. Visually, the Eta Carinae Nebula spans almost 2° in height and width and is incredibly complex. Close to its heart lies the star itself, but the eye sees the light filtering through the 20 arcsec-wide Homunculus Nebula surrounding the star; the star itself cannot be resolved. The integrated light from the whole swath of the nebula is about magnitude 4.8, but brightest near the center.

New Zealand observer Geoff Wingham obtained this superb four panel mosaic of the Eta Carina nebula using a Celestron 9.25-in. (235 mm) Schmidt-Cassegrain at *f*/5.2 and a one-shot color SBIG ST-2000XCM CCD on a Losmandy G-11 Mount. Each image was 6×300 s. Eta Carinae is the bright star in the centre.

The famous high resolution Hubble Space Telescope image of Eta Carinae itself showing the two expanding bipolar lobes of material thrown out in the 1843 explosion and now approaching a light-year in diameter. The nebulosity is known as the Homunculus Nebula. Image taken in 1996. Credit: J. Morse/K. Davidson/N. Smith/NASA.

Caldwell 93

NGC 6752
Starfish Globular Cluster in Pavo
Magnitude: 5.4
Size: 30′
Mag/Sec: 21
RA: 19 h 10 min 52 s
Dec: –59° 59′
Circumpolar: below 31°S
Starhop: 3¼°S and 9½°W from Peacock (α Pavonis)
Best Months: June to October
Distance: 13,000 light-years
Physical Size: ~110 light-years
Age: ~11 billion years
Best Visual Aperture: 10 cm and larger
Best Visual Filter: No filter advantage
CCD/DSLR Advice: Easy object with a ½° field
Celestial Neighbors: Galaxy NGC 6744 (C 101) lies 4°S
Miscellaneous: Surprisingly neglected naked-eye globular
Chart: See Caldwell 89

We enter Pavo the Peacock for Caldwell 93; the only stars brighter than fourth magnitude within, roughly 10°, are δ Pavonis (magnitude 3.5) and α Pavonis (magnitude 1.9, itself called the Peacock). The offsets in degrees of arc from this latter star are 3¼° south and 9½° west. Like Caldwell 86 this is another globular with relatively bright individual stars that can be held steady as separate suns in the eyepiece of a decent aperture telescope. OK, where you could see bucketloads of 12th and 13th magnitude stars with Caldwell 86, the same only applies to 13th and 14th magnitude stars here; this globular is 70% further away after all. One conspicuously bright star of magnitude 7.6 sits 4 arcmin south-southwest of the cluster's center. That star is GSC 90712141, but at 800 light-years it is a foreground Milky Way object, some 16 times nearer than the globular cluster; it is a double star though, with a magnitude 9.1 companion. Technically, NGC 6752 is one of the few naked-eye globular clusters, although whether an object is visible to the naked eye depends a lot on your sky background brightness, viewing experience, and the transparency. Nevertheless, only Omega Centauri, 47 Tucanae, and M22 are definitely brighter, although M4, M5, NGC 6397, and M13 are other fifth magnitude globulars only fractionally fainter. Not surprisingly, NGC 6752 is called a starfish simply because, to some observers, it looks a bit like one! The eye is good at seeing patterns in series of dots, whether they are there or not! The most concentrated part of NGC 6752 is the central 2 or 3 arcmin, but within that there is a smaller region, barely 20 arcsec across in which stars are crammed within a light-year or so of the globular's core. The Chandra X-ray observatory has detected six X-ray sources within this central region.

NGC 6752 from the Digitized Sky Survey. The field is 15 arcmin wide.

Caldwell 94

NGC 4755
The Jewel Box Cluster in Crux
Magnitude: 4.2
Size: 10′
Mag/Sec: Not relevant
RA: 12 h 53 min 36 s
Dec: –60° 21′
Circumpolar: below 30°S
Starhop: ¾°S and ¾°E from Becrux (ß Crux)
Best Months: March to June
Distance: 5,000 light-years
Physical Size: ~14 light-years
Age: ~15 million years
Best Visual Aperture: 10 cm and larger
Best Visual Filter: No filter needed
CCD/DSLR Advice: Easy object – great in color!
Celestial Neighbors: C98 (NGC 4609) & C99 (Coalsack)
Miscellaneous: Just like the name – a box of jewels

We have now entered the constellation of Crux, the Cross, a small and unmistakable constellation centered on–60° declination, which is invariably referred to as the Southern Cross. Compared with Cygnus, "the Northern Cross," it looks more like a small dagger, or even a kite, as the four bright stars form a deltoid shape. The easternmost star of the cross, Becrux (magnitude 1.3), is an easy marker to the "Jewel Box" cluster. From Becrux you just slide ¾° south and ¾° east.

To the naked eye this compact cluster can masquerade as a 4th magnitude star. With eagle eyesight in a black sky you may just sense two other stars curling away west and southwest from the Jewel Box, namely magnitude 5.7 DS Crucis and another magnitude 6.7 star half a degree from the cluster. During Sir John Herschel's four-year period in South Africa he used a massive 18-in. (45 cm) *f*/13 reflector at his residence "Feldhausen," near Claremont at the Cape of Good Hope, to study the southern hemisphere's wonders, including (of course) many future Caldwell objects. Herschel gave the Jewel Box his own catalog number HJ 3435 and commented about it as follows (with slight modification for his notebook/historic style of English): "The central star is extremely red; a most vivid and beautiful cluster of from 50 to 100 stars. Among the larger there are one or two evidently greenish; south of the red star is one of 13th magnitude, also red; and near it is one 12th magnitude, bluish... Several others laid down, of different shades of green."

Herschel got the attraction of this cluster spot on and identified why it is called the Jewel Box. In a compact 10 arcmin it has eight or nine stars of sixth to eighth magnitude that span a range of different colors along with many fainter colored offerings, too. Glorious!

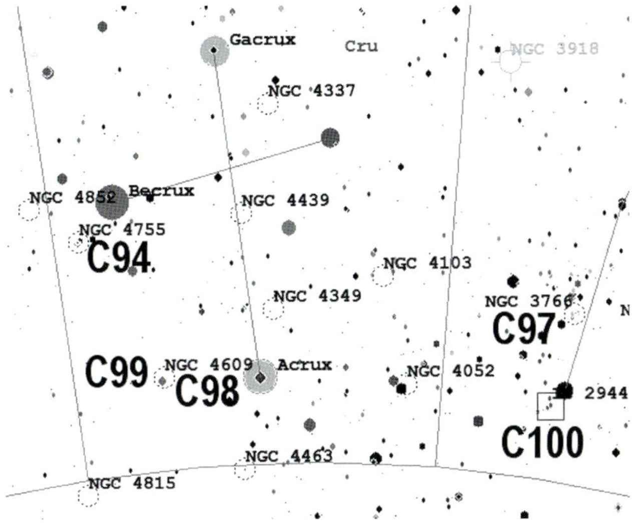

Caldwell objects 94, 98, and 99 in the Southern Cross, along with two other Caldwell objects, 97 and 100, are shown. This chart overlaps with the one on the Caldwell 91 page.

New Zealand observer Geoff Wingham obtained this image of NGC 4755 using a Celestron 9.25-in. (235 mm) Schmidt-Cassegrain at *f*/5.2 and a one-shot color SBIG ST-2000XCM CCD on a Losmandy G-11 Mount. The image is a median combined stack of more than 150 min of exposures.

Caldwell 95

NGC 6025
Open Cluster on Triangulum Australis/Norma border
Magnitude: 5.1
Size: 15′
Mag/Sec: Not relevant
RA: 16 h 03 min 18 s
Dec: –60° 25′
Circumpolar: below 30°S
Starhop: 3°N and 1°E from ß Triangulum Australis
Best Months: April to July
Distance: 2,500 light-years
Physical Size: ~11 light-years
Age: ~80 million years
Best Visual Aperture: Big binoculars and larger
Best Visual Filter: No filter needed
CCD/DSLR Advice: Easy object
Celestial Neighbors: C 89 (NGC 6025) lies 5°NE
Miscellaneous: Another constellation border-straddling cluster
Chart: See Caldwell 89

NGC 6025was mentioned six objects back, when discussing the number of open clusters in the constellation of Norma. This object lies right on the border of Triangulum Australis and Norma, with the center in the former constellation. To find your way there just find that Triangulum shape, which lies roughly 10° southeast of Alpha (α) Centauri/Rigel Kentaurus. The northernmost star in that triangle is magnitude 2.8 β Triangulum Australe. From there the cluster lies just 3° north and 1° east.

NGC 6025 is a rather loose collection of stars, with the brightest examples making the cluster stand out above the quite dense background star count in this region, on the edge of the Milky Way. The cluster resembles an X shape with two conspicuously bright stars to the southeast. These stars are the conspicuously bright MQ Triangulum Australis (magnitude 7.1) and a magnitude 8.1 star 1 arcmin to the south. MQ is the brightest cluster star, but in the X shape itself a dozen or so stars of between magnitude 8 and magnitude 10 keep the observer interested. As if to balance MQ Triangulum Australis on the southeastern edge a fainter variable star, nominally magnitude 9.3, sits on the northwestern edge in Norma; this is V350 Normae. One arcminute to its west lies a brighter star of magnitude 8.3. A number of faint galaxies surround this cluster, and one lies within its boundaries, but these are only targets for the CCD imager. With north at the top the galaxy PGC 2793246 (magnitude 15.4) lies at the 7'o-clock position, PGC 2793236 (magnitude 16.3) lies at the 4'o-clock position, and PGC 2793235 (magnitude 16.7) lies at the 1'o-clock position. All these galaxies are a few arcminutes outside the cluster boundary, but 2 arcmin *within* the northern boundary you may be able to record magnitude 15.3 PGC 56851, just 3 arcmin southeast of the aforementioned V350 Normae. This galaxy lies absolutely smack bang on the Norma/Triangulum Australe border.

NGC 6025 from the Digitized Sky Survey. The field is 15 arcmin wide.

Caldwell 96

NGC 2516
Open Cluster in Carina
Magnitude: 3.8
Size: 25′
Mag/Sec: Not relevant
RA: 07 h 58 min 00 s
Dec: –60° 45′
Circumpolar: below 30°S
Starhop: 1¼°S and 3°W from Avior (ε Carinae)
Best Months: December to March
Distance: 1,300 light-years
Physical Size: ~9 light-years
Age: ~110 million years
Best Visual Aperture: Big binoculars give a great view
Best Visual Filter: No filter needed
CCD/DSLR Advice: Easy object
Celestial Neighbors: Relatively barren surroundings
Miscellaneous: The southern hemisphere's Beehive
Chart: See Caldwell 74

We are back in Carina the Keel for our next open cluster, our fourth trip to this constellation. Western and central Carina are very barren regions for the open cluster observer, but there are eleven good ones crammed in the northeastern corner (such as the Caldwell 91 Pincushion). However, away from that corner, NGC 2516 steals the show in central Carina, with no other significant deep sky objects anywhere near it. Locate the magnitude 2.2 star Avior (ε Carinae), and the cluster we seek is just 1¼° south and 3° west of that star. Some observers refer to Caldwell 96 as the southern Beehive, claiming a similarity to the "real" Beehive cluster, M44 in Cancer (sometimes called Praesepe). However, there are considerable differences, not least M44's diameter of some 95 arcmin compared with C96's diameter of 25 arcmin or so (there is some dispute about NGC 2516's extent). Certainly, NGC 2516 is much more densely packed with bright stars within a small 25 arcmin diameter. The brightest star within this circle is V374 Carinae in the southeast corner which has a nominal magnitude of 5.8. On the very northeastern edge of the cluster boundary is another bright star with some variability, V460 Carinae, nominally magnitude 5.2. Just beyond the northwestern cluster edge a magnitude 5.7 star catches the eye too. A further 15 cluster stars brighter than magnitude 10 are suspected of variability. The catalog magnitude of 3.8 for NGC 2516 sounds quite impressive, but the cluster would be more dazzling if it were made up from, say, 100 stars of magnitude 8.8; then it would start looking like the Pincushion. Instead, the cluster's light is dominated by a dozen stars from 5th to 8th magnitude, which makes it a pretty and compact cluster but not a stunner.

NGC 2516 from the Digitized Sky Survey. The field is 30 arcmin wide.

Caldwell 97

NGC 3766
Open Cluster in Centaurus
Magnitude: 5.3
Size: 15′
Mag/Sec: Not relevant
RA: 11 h 36 min 12 s
Dec: –61° 37′
Circumpolar: below 29°S
Starhop: 1½° due N from λ Centauri
Best Months: March to June
Distance: 5,800 light-years
Physical Size: ~25 light-years
Age: ~15 million years
Best Visual Aperture: 10 cm and larger
Best Visual Filter: No filter needed
CCD/DSLR Advice: Easy object
Celestial Neighbors: Collinder 249 (C 100), lies 2°S
Miscellaneous: Fine cluster in an awesome constellation
Chart: See Caldwell 91 or Caldwell 94

The most frustrating aspect of the Caldwell catalog is its rigid adherence to listing objects in descending declination order (ignoring the two slight errors in this regard). Because of this, neighboring objects, sometimes in the same constellation, can be many catalog numbers apart because other objects at intermediate declinations, but far removed in right ascension have squeezed into the declination gap. Thus, Caldwell 97/NGC 3766 is an object that you will want to observe alongside the Jewel Box (C 94), the Pincushion (C91), and the Eta Carinae Nebula (C92).

NGC 3766 is in the narrow corridor of Centaurus that is squeezed between Crux and northeastern Carina. A short hop of 1½° north from magnitude 3.1 λ Centauri will get you to the object. λ Centauri is just 5° west of Acrux, the bright southernmost star in the cross. Caldwell 97 is a much tighter bunch of stars than the previous Caldwell object. It only covers a quarter of a degree, but despite its dimmer integrated magnitude of 5.3 it is an equally pretty object; it simply does not have the fifth and sixth magnitude stars that boost Caldwell 96 into a third magnitude category. Three stars of about fifth magnitude, in a 2° line, do lead you into the cluster from the southeast, starting with V810 Centauri.

The brightest stars in Caldwell 97 are eight in number and are of seventh and eighth magnitude. Two dozen more down to tenth magnitude make the cluster sparkle, although they look mainly white in color. An additional two dozen add a background carpet of fainter stars to the view. Just above the cluster's northern edge lies the eclipsing binary star BF Centauri, whose maximum and minimum magnitudes are 8.5 and 9.4. Lying literally on the cluster's visible edge and situated roughly midway (2,300 light-years) between Caldwell 97 and Earth is the luminous 500 sun-power star V910 Centauri, which varies between magnitude 7.2 and 7.3.

NGC 3766 from the Digitized Sky Survey. The field is 15 arcmin wide.

Caldwell's 98 and 99

NGC 4609 Coalsack Cluster in Crux and
Coalsack Dark Nebula in Crux, Centaurus, and Musca
Magnitude: 6.9 (cluster)
Size: 6′ (cluster); 7° × 5° (Coalsack)
Mag/Sec: Not relevant
RA: 12 h 42 min 18 s (cluster) 12 h 53 min (Coalsack)
Dec: –63° 00′ for both
Circumpolar: below 27°S
Starhop: Cluster is 2°E from Acrux (α Crux)
Coalsack center is ~3½° from Acrux
Best Months: March to June
Distance: 4,200 light-years (cluster) and 600/800 light-years (dark nebulae)
Physical Size: ~7 light-years (cluster) and ~100 light-years (Coalsack)
Age: 80 million years for the cluster
Best Visual Aperture: Big binoculars work for both
Best Visual Filter: No filters required
CCD/DSLR Advice: Try DSLR + 100 mm lens for Coalsack
Celestial Neighbors: NGC 4755 (C94 Jewel Box) nr Becrux
Miscellaneous: C99 is a "must see" on any trip south
Chart: See Caldwell 94

Caldwell objects 98 and 99 are two completely different types of object at completely different distances from Earth. However, they are consecutive objects in Patrick's catalog and occupy the same part of the constellation Crux (C99 actually spans three constellations), so it is logical to deal with them together. Once again the bright star Acrux is the marker to the objects of interest. The magnitude 6.9 open cluster NGC 4609 lies at an estimated 4,200 light-years from Earth, whereas the dark nebula known as the Coalsack is thought to consist of two overlapping nebulae roughly 600 and 800 light-years distant. The cluster is just 2° east and a fraction north of Acrux, whereas the dark nebulae span a large region roughly 7° high by 5° wide just east of Acrux and with the star cluster near to its western edge. You may spot a possible anomaly at this point.

If the star cluster is roughly six times further away than the dark, dusty Coalsack and yet lives within its borders, how can we see it at all? In fact, professional astronomers have analyzed the light from the stars making up the Coalsack cluster and the stars are considerably dimmed and reddened by the intervening dust. Although the Coalsack itself is obvious to the naked eye as a dark, obscuring, and eerie patch, it is not opaque. The stars in a much fainter cluster, Hogg 15, 10 arcmin southeast of NGC 4609, are also dimmed by the Coalsack. The two clusters are not associated; they lie either side of the fifth magnitude star BZ Crucis. Tiny 6 arcmin diameter NGC 4609 itself reveals twenty stars of magnitude 13 or brighter in a telescope. Its three brightest offerings are all magnitude 9.6, so this is not a dramatic cluster, but it is a visual novelty shining, as it does, through the dark veil of the Coalsack.

NGC 4609 from the Digitized Sky Survey. The field is 15 arcmin wide.

A 2004 photograph of the Coalsack and the Southern Cross taken from Hakos, Namibia, using a 135 mm *f*/4.0 lens, a 20 min exposure and Fuji Provia 400F film. The field is approximately 10° wide. Image by Till Credner of AlltheSky.com.

Caldwell 100

Collinder 249
Open Cluster in Centaurus
Magnitude: 2.9 (including nebulosity)
Size: 65′ × 40′ (including nebulosity)
Mag/Sec: Not relevant
RA: 11 h 38 min 18 s
Dec: –63° 22′
Circumpolar: below 27°S
Starhop: immediately SE of λ (lambda) Centauri
Best Months: March to June
Distance: 6,000 light-years
Physical Size: ~100 light-years
Age: ~7 million years
Best Visual Aperture: Big binoculars are sufficient
Best Visual Filter: No filter needed
CCD/DSLR Advice: H-alpha filtering enhances nebulosity
Celestial Neighbors: NGC 3766 (C 97) lies 2°N
Miscellaneous: Loads to image, but less to see
Chart: See Caldwell 91 or Caldwell 94

We will gloss over the much debated mystery regarding the true identity of Caldwell 100, which appears to be a complex combination of a Patrick typo (mixing the Greek letters γ and λ) and confusion between IC 2944, IC 2948, and Collinder 249. The majority opinion is that Caldwell 100 is Collinder 249, the open cluster of stars half a degree southeast of magnitude 3.1 λ Centauri, plus a lot of the surrounding region and its nebulosity. A half dozen stars of seventh or eighth magnitude are the main features of this cluster, but various star catalogs and planetarium software seem hopelessly confused as to what catalog designations refer to which objects. We were in this region recently when we used this star to hop north to Caldwell 97, the open cluster NGC 3766. The cluster we seek is immersed in the nebulosity of IC 2948, and a degree-wide CCD field with λ Centauri placed near the northwest corner will record this nebulosity well, especially with deep exposures through H-alpha filters. The cluster itself is mainly a half degree streak, from northwest to southeast, centered on magnitude 6.5 V871 Centauri. The Collinder/Lundmark magnitude of 2.9 is not really relevant to the cluster, as it seems to include a huge amount of nebulosity out to a size of 65′ × 40′. Determining just how many stars make up this loose-trailing aggregation seen through a telescope eyepiece is rather tricky. Including V871 Centauri there are more than a dozen stars down to magnitude 9. Beyond this limit the fainter stars tend to merge with the background. So despite having the prestigious Caldwell designation of 100 this object or objects is rather disappointing visually, even if it can look good in a deep H-alpha image such as the close up by Daniel Verschatse shown on the opposite page. Caldwell 100 is perhaps more famous for its controversial identity than its visual appearance!

An enlargement of the central region of a splendid H-alpha image of the Caldwell 100 (IC 2948/Collinder 249) region taken by Daniel Verschatse at his Observatorio Antilhue, San Esteban, Chile. Astro-Physics AP155EDFS @ *f*/7.2 (prime focus). Astro-Physics AP1200-GTO mount. Custom Scientific 4.5 nm Hα filter. SBIG ST-10XE self-guided CCD. 12 × 15 min @ −15°C. 2003 April 2 @ 02:30 UTC. © Daniel Verschatse – Observatorio Antilhue – Chile.

Caldwell 101

NGC 6744
Galaxy in Pavo
Magnitude: 8.6
Size: 21′ × 13′
Mag/Sec: 23
RA: 19 h 09 min 42 s
Dec: −63° 51′
Circumpolar: Below 27°S
Starhop: 7°S and 9½°W from Peacock (α Pavonis)
Best Months: June to October
Distance: 35 million light-years
Physical Size: ~210,000 light-years
Age: Billions of years
Best Visual Aperture: 15 cm and larger is recommended
Best Visual Filter: Deep sky filter may help
CCD/DSLR Advice: Potential stunner with LRGB
Celestial Neighbors: Globular NGC 6752 (C 93) lies 4°N
Miscellaneous: Host to Supernova 2005 discovered by Martin and Monard
Chart: See Caldwell 89

Caldwell 101 is best looked at on the same night as Caldwell 93 because the former object (C 101) is just 4° due south of the latter, but we have stuck with Patrick's sequence throughout this book and finally arrived at this splendid galaxy, better known as NGC 6744. Sadly, this is the last galaxy in the Caldwell catalog. As a CCD imager, you might have stuck far more galaxies in the catalog, but Patrick is primarily a visual observer.

NGC 6744 is a splendid galaxy, located in the rather barren (for deep sky observers) central region of Pavo the Peacock. Indeed, although it is a long haul, the bright star α Pavonis, also called the Peacock, is arguably the best steppingstone to this wonderful spiral. From that star the hop is 7° south and 9½° west. The best amateur CCD images of this galaxy reveal it is one of the most beautiful in the night sky with a barred spiral form and exquisitely intricate spiral arms, and yet how many northern hemisphere observers have even heard of it? It does not even have a nickname! Yet it is an absolute stunner. It is a huge galaxy, too, with a diameter in excess of 200,000 light-years – not quite as large as Caldwell 35, but that object is 300 million light-years distant, whereas Caldwell 101 is just 35 million light-years away. As a consequence and with its sheer beauty, it is, arguably, the most awesome super-large galaxy that any amateur can image; any southern hemisphere amateur that is. Unfortunately NGC 6744 is not stunning visually. A quick comparison of its very large size, but modest magnitude of 8.6, tells us it has a surface brightness of around magnitude 23/sq. arcsec. In practice this means that the galaxy may only appear a third or a quarter as large in the telescope eyepiece, and even in a big Dobsonian the spiral arms are a real challenge even from a dark site.

A mouthwatering image of the beautiful galaxy NGC 6744 by the deep sky imaging master Robert Gendler. This remotely acquired image was taken with a 14.5-in. (37 cm) *f*/9 RCOS Ritchey-Chrétien with field flattener, based in Western Australia. Paramount ME mount and SBIG STL-11000XM CCD. © Robert Gendler.

Caldwell 102

IC 2602
Theta Carinae Cluster
Magnitude: 2.0
Size: 100′
Mag/Sec: Not relevant
RA: 10 h 42 min 54 s
Dec: −64° 24′
Circumpolar: Below 26°S
Starhop: 2¾°S and 1¼°E from PP Carinae
Best Months: February to May
Distance: 500 light-years
Physical Size: ~15 light-years
Age: ~50 million years
Best Visual Aperture: Binoculars are sufficient
Best Visual Filter: No filter needed
CCD/DSLR Advice: Easy with DSLR + 500-mm lens
Celestial Neighbors: C91, C92 (to N); Melotte 101 (to S)
Miscellaneous: Not noted by J. Herschel
Chart: See Caldwell 91

The theta Carinae cluster is the last open cluster in the Caldwell catalog. From here on we have just a nebula, a planetary nebula, and another bucketload of globular clusters. We are back in that northeastern corner of Carina the Keel again; you must know it pretty well by now! We have already found the Pincushion, Eta Carinae, and, just across the border, NGC 3766 in this region. From PP Carinae it is a hop 2¾° south and 1¼° east. IC 2602 is a big open cluster and has even been referred to as the southern Pleiades on occasion. However, the Pleiades has six stars between magnitude 2.8 and 4.3 in a distinctive "wheelbarrow" shape, whereas IC 2602 has a single central star of magnitude 2.7, but the five next brightest are only between magnitude 4.8 and 5.2 and the shape is not distinctive. However, it is a pretty grouping, and, as part of the other marvelous spectacles in this region, it makes northeastern Carina a deep sky heaven. The magnitude 2.7 central star of Caldwell 102 is bright enough to have a Greek letter designation; it has the designation θ Carinae. Right on the southern boundary of this large cluster (inside the boundary according to some definitions of the diameter of IC 2602) is a fainter star cluster called Melotte 101. It is just 40 arcmin south of θ Carinae, and its center is 15 arcmin east of the fifth magnitude star V364 Carinae. Melotte 101 is 13 arcmin across and has a total magnitude of 8.0 made up from three dozen stars of around twelfth magnitude and two of tenth magnitude. At 6,800 light-years it is 14 times more distant than Caldwell 102 itself and has an age of about 60 million light-years and a diameter of roughly 25 light-years. So if Caldwell 102 was removed to the distance of Melotte 101 it would only look 60% as wide as that cluster. As it is the more distant object appears to shine through the extreme outer edge of the much nearer one.

IC 2602 from the Digitized Sky Survey. The field is 15 arcmin wide.

Caldwell 103

NGC 2070
Tarantula Nebula in Dorado
Magnitude: 4
Size: 30′ × 20′
Mag/Sec: average ~20
RA: 05 h 38 min 36 s
Dec: –69° 05′
Circumpolar: Below 21°S
Starhop: 3½°S and 0.6°W from δ Doradus magnitude 4.3 or 16½°S and 5½°W from Canopus (α Carinae)
Best Months: November to February
Distance: 170,000 light-years
Physical Size: ~1500 light-years
Age: Central cluster ~3 million years
Best Visual Aperture: 10 cm and larger
Best Visual Filter: UHC filter enhances the view
CCD/DSLR Advice: Superb LRGB and Ha target
Celestial neighbours: Many splendors in the LMC
Miscellaneous: Top drawer deep sky wonder

For most amateur astronomers born prior to, say, 1970 the Tarantula Nebula is a deep sky object most often associated with Supernova 1987A, the only supernova in our Milky Way's vicinity to have been seen with the naked eye in the past 400 years. The Tarantula is a unique object in this Caldwell catalog – a bright and large nebula, easily visible to the naked eye and yet it is outside our own galaxy. It lives in the Large Magellanic Cloud (LMC) some 170,000 light-years away from our Milky Way location, a distance slightly smaller than the diameter of that awesome galaxy we recently visited, Caldwell 101. The LMC is, of course, along with the smaller SMC, a companion galaxy to the Milky Way and not a truly independent island universe. Nevertheless, for an object to span half a degree at that distance implies a diameter of some 1,500 light-years, and that is just the inner portion visible to amateur astronomers. The Tarantula's full extent is twice as large.

The LMC is a very easy object to locate, as it looks just like a ghostly cloud 11° × 9° in extent in southern Dorado, stretching south across the border into northern Mensa. The Tarantula itself lies just one degree above (north) of that Dorado/Mensa border near the southeastern edge of the LMC. Frankly, despite many other wondrous sights in that area, it is unmistakable in a modest telescope, as it does actually look a bit like a Tarantula. The field is adorned with whirls and loops and knots. This nebula is the largest HII region known in the entire local group of galaxies. The Tarantula contains, at its heart, a magnitude 8.3 star cluster some 5 arcmin across. The two objects, nebula and cluster, are sometimes referred to as 30 Doradus. The Tarantula Nebula was originally thought to be a star, but in 1751 Abbé Nicolas Louis de Lacaille discovered its true nebular nature, and he included it in his 1755 catalog.

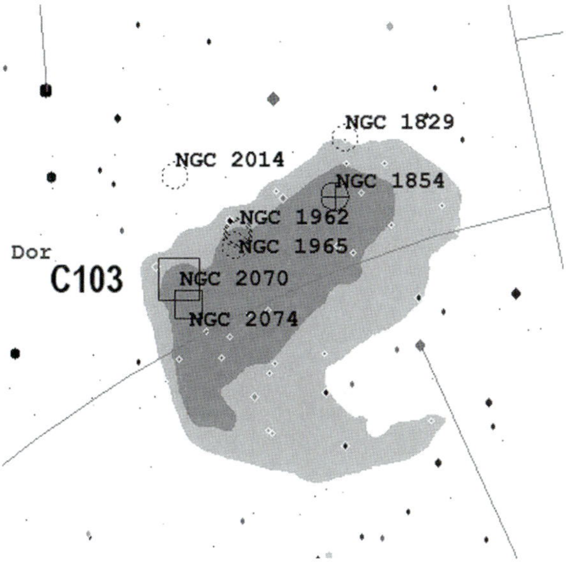

The Large Magellanic Cloud spans Dorado and Mensa and the location of the Tarantula nebula, Caldwell 103, with respect to the LMC is shown.

New Zealand observer Geoff Wingham obtained this image of the Tarantula nebula in the LMC using a Celestron 9.25-in. (235 mm) Schmidt-Cassegrain at *f*/5 and a one-shot color SBIG ST-2000XCM CCD on a Losmandy G-11 Mount. The image is a median combined stack of more than 200 min of exposures.

Caldwell 104

NGC 362
Globular Cluster in Tucana
Magnitude: 6.4
Size: 14′
Mag/Sec: 21
RA: 01 h 03 min 14 s
Dec: −70° 51′
Circumpolar: Below 19°S
Starhop: 5½°N and 2¾°E from ß Hydri
Best Months: All year from Australian latitudes
Distance: 28,000 light-years
Physical Size: ~115 light-years
Age: ~10 billion years
Best Visual Aperture: 10 cm works, but larger is better
Best Visual Filter: No filter advantage
CCD/DSLR Advice: Easy to image in mono or color
Celestial Neighbors: SMC and 47 Tucanae
Miscellaneous: Overshadowed by 47 Tucanae

Do you like globular clusters? Good! Because five of the final six Caldwell objects are globulars, and there is one absolute stunner. But we start more modestly with magnitude 6.4 NGC 362 that can be found just to the north of the Small Magellanic Cloud (SMC). Locating the SMC is easy; it lives in the southeastern corner of Tucana roughly 4° north of magnitude 2.8 β Hydri. From that star the globular cluster NGC 362 is 5½° to the north and 2¾° to the east. The awesome globular 47 Tucanae (NGC 104/Caldwell 106) is also close to the SMC (to the east) and with a wide 5-degree CCD/DSLR shot you can fit the three objects in the same frame. With a full frame (24 × 36 mm) DSLR sensor a 250 mm focal length lens is about optimum for the SMC/NGC362/NGC 104 trio. NGC 362 is a rather neglected globular cluster and not just because it is so very far into the southern hemisphere. The problem is that the SMC itself is a big attraction, replete with open clusters of its own, and for anyone who loves staring at these objects the stunning 47 Tucanae is just 3° away, ten times brighter, and four times wider. The majority of the stars in NGC 362 are roughly 15th or 16th magnitude and so only the largest amateur telescopes will allow you to hold individual stars steady across the disc. Through telescopes of 10 or 15 cm aperture the cluster will look a quarter the size of its catalog diameter, but with a noticeably stellar core. A handful of stars in the cluster are 12th or 13th magnitude.

When viewing the trio of two globulars and the SMC together it is interesting to know the relative distances of the objects, as it is easy to think of them as lying at roughly the same distance; after all they are all satellites of the Milky Way. The SMC itself lies at a distance of roughly 200,000 light-years, far more distant than the globular clusters. By comparison NGC 362 is 7× closer and 47 Tucanae is 14× closer, an indication of just how large the SMC really is.

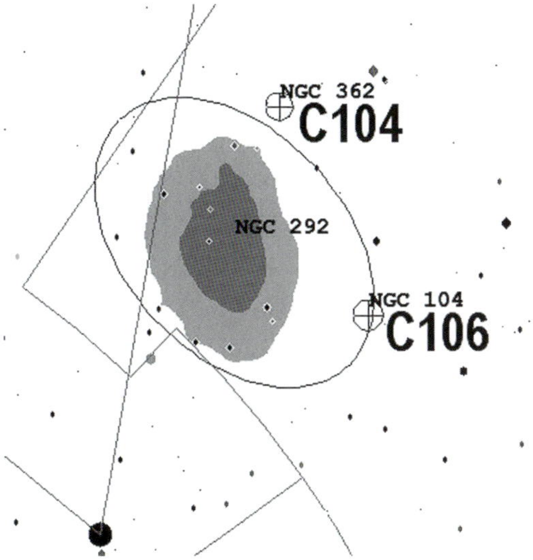

The chart shows the location of Caldwell's 104 and 106 with respect to the Small Magellanic Cloud (SMC).

A splendid image of NGC 362 taken by Daniel Verschatse at his Observatorio Antilhue, San Esteban, Chile. RCOS 14.5-in. (37cm) Ritchey-Chrétien @ *f*/9 (prime focus). Astro-Physics AP1200-GTO mount. SBIG STL-11000 self-guided CCD. LRGB 60:20:20:20 min @ –20°C. 2006 Aug 20 @ 09:00 UTC. © Daniel Verschatse - Observatorio Antilhue – Chile.

Caldwell 105

NGC 4833
Globular cluster in Musca
Magnitude: 7.0
Size: 13′
Mag/Sec: 21
RA: 12 h 59 min 35 s
Dec: –70° 52′
Circumpolar: Below 20°S
Starhop: 1¾°S and 2°E from α Muscae
Best Months: All year from Australian latitudes
Distance: 20,000 light-years
Physical Size: ~75 light-years
Age: ~11 billion years
Best Visual Aperture: 10 cm and larger
Best Visual Filter: No filter advantage
CCD/DSLR Advice: Easy to image in mono or color
Celestial Neighbors: Globular NGC 4372 (C108)
Miscellaneous: Blue magnitude 8 foreground star contrasts nicely with "gold" globular

Draw a line from the previous Caldwell globular, through the southern pole and out the other side, and you come to NGC 4833, the second of our "five in a row" globulars. From any southern hemisphere locations south of 20° latitude both these globulars are circumpolar, that is, they never set. But, because they are diametrically opposite the southern pole, they are at their best 6 months apart. NGC 4833 lives in Musca, just 1¾° south and 2° east of that constellation's brightest star, magnitude 2.7 α Muscae. Alternatively, if you can find magnitude 3.6 δ Muscae, the globular is a mere 40 arcmin to its NNW. Three Caldwell's down the line, Caldwell 108, another globular, is just 3° to the southwest of Caldwell 105, yet another irritating sequence anomaly caused by the catalog's rigid adherence to a descending declination order! As with many far southern hemisphere objects there seems to be a range of magnitudes and diameters quoted for this globular, and through a telescope it may only look half the diameter quoted. Like the previous Caldwell globular the individual stars in NGC 4833 are 15th or 16th magnitude, and the brightest stars (apart from one foreground star of magnitude 8.8, 2.5 arcmin north of the nucleus) are only 12th magnitude. Well, there is one of magnitude 11 and a couple of dozen of magnitude 12. So this is another globular cluster that will not sparkle except in a massive telescope. Stephen O'Meara has nicknamed this the Southern Butterfly globular as, visually, butterfly wings or petals seem to appear caused by the chains of stars to the east and west of the object. Anyone observing this object will be compelled to compare it with the aforementioned nearby globular Caldwell 108, otherwise known as NGC 4372.

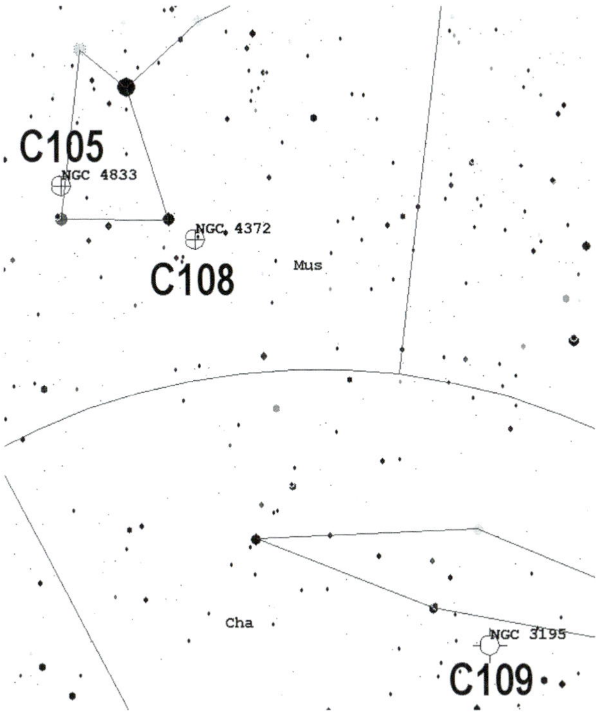

The chart shows the two Caldwell globular clusters in Musca (C105 and C108), as well as the final, 109th Caldwell object, the planetary nebula NGC 3195.

A splendid image of NGC 4833 taken by Daniel Verschatse at his Observatorio Antilhue, San Esteban, Chile. RCOS 14.5-in. (37 cm) Ritchey-Chrétien @ *f*/9 (prime focus). Astro-Physics AP1200-GTO mount. SBIG STL-11000 self-guided CCD. LRGB 50:12:12:12 min @ –20°C. 2005 April 11 @ 04:30 UTC. © Daniel Verschatse - Observatorio Antilhue – Chile.

Caldwell 106

NGC 104
Globular Cluster 47 Tucanae
Magnitude: 4
Size: 50′
Mag/Sec: 21
RA: 00 h 24 min 05 s
Dec: –72° 05′
Circumpolar: Below 18°S
Starhop: 5° due N from ß Hyi
Best Months: All year from Australian latitudes
Distance: 15,000 light-years
Physical Size: ~215 light-years
Age: ~10 billion years
Best Visual Aperture: Awesome with any optical aid
Best Visual Filter: No filter advantage
CCD/DSLR Advice: Awesome target in mono or color
Celestial Neighbors: SMC and Globular NGC 362 (C104)
Miscellaneous: A jaw dropping "WOW!" in any telescope
Chart: See Caldwell 104

Ask any northern hemisphere amateur astronomer to name a few globular clusters in the southern celestial sky, and the chances are the vast majority will name but two: Omega Centauri (Caldwell 80/NGC 5139) and this one, 47 Tucanae. They are both awesome, but how do they compare? Well, both are close to magnitude 4 (Omega Centauri is slightly brighter), 50–55 arcmin in diameter (again Omega is slightly superior) and as a result they have a similar surface brightness. Both objects show myriads of stars as bright as 13th, 12th, and even 11th magnitude, although 47 Tucanae's slightly closer distance to Earth arguably gives it the edge in the sparkling jewels category. Although a few smaller globulars that are close to us have average star magnitudes that are brighter than either of these two monsters it is the sheer quantity of sparkling stars across an almost one degree field that generates the "WOW!" factor. 47 Tucanae is 215 light-years across compared to Omega Centauri's record-breaking (for our Milky-Way's globulars) 280 light-years. In terms of brightness profile Omega Centauri is less concentrated in the core than 47 Tucanae, but the sheer number of stars in the former object is awesome, at least 3 million stars, compared to just one million in 47 Tucanae. However, 47 Tucanae's combination of a dense core and less "flat" appearance make it hang in the eyepiece like a genuine three-dimensional sphere. At its core live 15,000 stars packed together so tightly that they occasionally collide, forming "blue straggler" stars from the merger. Deep sky observing would be impossible from the core of this object. Actually, the biggest difference between these objects may have nothing to do with their appearance. The simple fact is that Caldwell 80 lives at declination –47°, and Caldwell 106 lives at –72°. From southern Europe and the southern United States you can see the former but not the latter.

A stunning image of the remarkable globular cluster 47 Tucanae by the deep sky imaging master Robert Gendler. This remotely acquired image was taken with a 14.5-in. (37 cm) *f*/9 RCOS Ritchey-Chrétien with field flattener, based in Western Australia. Paramount ME mount and SBIG STL-11000XM CCD. This is a two frame mosaic assembled from a total of eight hours of imaging data. © Robert Gendler.

Caldwell 107

NGC 6101
Globular Cluster in Apus
Magnitude: 9.2
Size: 11′
Mag/Sec: 23
RA: 16 h 25 min 50 s
Dec: −72° 12′
Circumpolar: Below 18°S
Starhop: 3°S and 2°W of α Triangulum Australe
Best Months: All year from Australian latitudes
Distance: 50,000 light-years
Physical Size: ~160 light-years
Age: ~12 billion years
Best Visual Aperture: 15 cm and larger
Best Visual Filter: No filter advantage
CCD/DSLR Advice: Easy to image in mono or color
Celestial Neighbors: Relatively barren surroundings
Miscellaneous: Run of the mill faint globular
Chart: See Caldwell 89

This is our one and only visit to the constellation of Apus, the Bird of Paradise. Unfortunately, the region is not a paradise for deep sky observers or lovers of bright stars. The brightest stellar offerings are the magnitude 3.8 stars α and γ Apus, but fortunately, magnitude 1.9 α Triangulum Australis is just over the border and nearer to the globular listed as Caldwell 107. A hop of 3° south and 2° west from α TrA will get you there. The only other globular in Apus is IC 4449, which is a dismal magnitude 10.6 offering in the extreme south of the constellation at a declination of 82° south.

Our quarry NGC 6101 will not dazzle you at the eyepiece, either; at magnitude 9.2 it is more than 100 times fainter than Omega Centauri or 47 Tucanae, but it does have some attractions if you have access to a very large amateur telescope or a CCD camera. Although a handful of the brightest stars in this globular cluster shine at 12th or 13th magnitude they are a rarity, and for significant numbers you have to be able to reach 14th and 15th magnitude. Even then only a score of stars are resolvable, and unless you can reach 16th or 17th magnitude NGC 6101 will look just like a fuzzy object, denser in the center and with hints of a few east-west chains. This globular cluster is quite a large one, with a physical diameter of some 160 light-years; the problem is, at 50,000 light-years from Earth it is more than three times further than 47 Tucanae which, all things being equal, will result in it being 9 times dimmer (2.4 magnitudes). NGC 6101 has, like many globular clusters, a significant blue straggler stellar population, and in this case there appear to be many heavy examples concentrated in the core, suggesting that there are some close and massive binary star pairs in the cluster and many examples of merged massive stars.

NGC 6101 from the Digitized Sky Survey. The field is 15 arcmin wide.

Caldwell 108

NGC 4372
Globular Cluster in Musca
Magnitude: 7.2
Size: 19′
Mag/Sec: 22
RA: 12 h 25 min 45 s
Dec: –72° 40′
Circumpolar: Below 18°S
Starhop: ½°S and ½°W from γ Muscae
Best Months: All year from Australian latitudes
Distance: 19,000 light-years
Physical Size: ~105 light-years
Age: ~12 billion years
Best Visual Aperture: 10 cm and larger
Best Visual Filter: No filter advantage
CCD/DSLR Advice: Easy to image in mono or color
Celestial Neighbors: Globular NGC 4833 (C 105)
Miscellaneous: Blue magnitude 6 foreground star contrasts nicely but dazzles
Chart: See Caldwell 105

We were in Musca only three objects ago, looking at its other globular cluster, NGC 4833, 3° further northeast. You were warned we would be back here soon, and here we are! So, how does this second offering in Musca stack up? Well, this new globular is fractionally closer to us and, physically, some 40% larger. But despite this, it is no brighter than its neighbor, in fact, a touch fainter, and so its surface brightness per square-arcsecond is a full magnitude fainter. It is easy to find because it sits only half a degree south and half a degree west of magnitude 3.8 γ Muscae.

Only half a degree to the globular's southeast you can find a distinctive small triangle of stars with magnitudes 5.9, 7.0, and 7.1. Just 5 arcmin northwest of the globular cluster's core lies the magnitude 6.6 star GSC 92362380, which rather complicates any visual inspection of the cluster through a telescope. This star is, of course, a foreground object; it lies just 300 light-years from Earth, or 63 times closer than the globular itself.

The clue to NGC 4372's relatively faint appearance (and there is some disagreement about its magnitude and size) is that it is a very open globular cluster and not as densely packed as some we have encountered. Of course, deciding where the outer edges of a globular cluster fade into the foreground carpet of Milky Way stars is always a problem unless each star in the region has its distance determined accurately. In a small finder telescope the concentrated light from the magnitude 6.6 star may well blind you to the fainter globular spread over the claimed 19 arcmin of sky, unless you are a practiced expert with using averted vision. Despite its distance NGC 4372 does have a reasonable number of 14th magnitude stars along with a few 13th and even 12th magnitude examples.

A splendid image of NGC 4372 taken by Daniel Verschatse at his Observatorio Antilhue, San Esteban, Chile. RCOS 14.5-in. (37 cm) Ritchey-Chrétien @ *f*/9 (prime focus). Astro-Physics AP1200-GTO mount. SBIG ST-11000 self-guided CCD. LRGB 30:9:9:9 min (individual exposures 3 min) @ −20°C. 2005 April 3 @ 05:00 UTC. © Daniel Verschatse – Observatorio Antilhue – Chile.

Caldwell 109

NGC 3195
Planetary Nebula in Chamaeleon
Magnitude: 11.6
Size: 38″
Mag/Sec: 19
RA: 10 h 09 min 21 s
Dec: –80° 52′
Circumpolar: Below 10°S
Starhop: 11°S and 4°E of Miaplacidus (ß Carinae)
Best Months: All year from Australian latitudes
Distance: 5,500 light-years
Physical Size: ~1 light-year
Age: ~5,000 years
Best Visual Aperture: 30 cm and larger
Best Visual Filter: OIII or UHC
CCD/DSLR Advice: Tiny object, but high Dec. aids tracking
Celestial Neighbors: Very barren polar region
Miscellaneous: Disappointing or invisible in small apertures
Chart: See Caldwell 105

Patrick's final choice for his Caldwell catalog is a planetary nebula in Chamaeleon at a declination of –80° 52′. There are more southerly deep sky objects, such as the globular cluster IC 4449 in Apus and the open cluster NGC 1841 in Mensa, but this planetary nebula seems a reasonable place to stop. We have seen far too many globulars in the last few pages! With no stars brighter than magnitude 4.0 the Chamaeleon is not the most exciting constellation. The planetary nebula we seek is situated 2° west and 20 arcmin south of magnitude 4.4 δ Chamaeleontis, a star that is quite recognizable in a finder because of its magnitude 5.6 companion 5 arcmin to the north. If you are brave enough for a long hop from a much brighter star, the object is 11° south and 4° east of Miaplacidus (ß Carinae), a magnitude 1.7 star in the Southern Keel. The first thing to bear in mind with this planetary nebula is that it is a faint one. At only magnitude 11.6 and with a surface brightness of magnitude 19/sq. arcsec this will not blow your mind. It is a nice challenge for the deep sky imager, though, as at a declination of –80°, if you are accurately polar aligned, the tracking errors will be reduced by a factor of almost six times compared to an object on the celestial equator where the sky moves at the full 15 arcsec per second of time. On the other side of the Caldwell sky, near the north polar region, Caldwell 2 (the Bow-Tie) is actually 0.7 magnitudes fainter, roughly half as bright. However, in addition to that northern object's nebulosity it has a central star of roughly equal brightness. Caldwell 109's central star is a feeble magnitude 15.3. Nevertheless, NGC 3195 has a higher surface brightness than most deep sky objects and is easily spotted in a modest aperture telescope as a bloated star at magnifications above 100×. An OIII filter and a large aperture will enable you to view this object as a small disk and (with a vivid imagination) some inner structure.

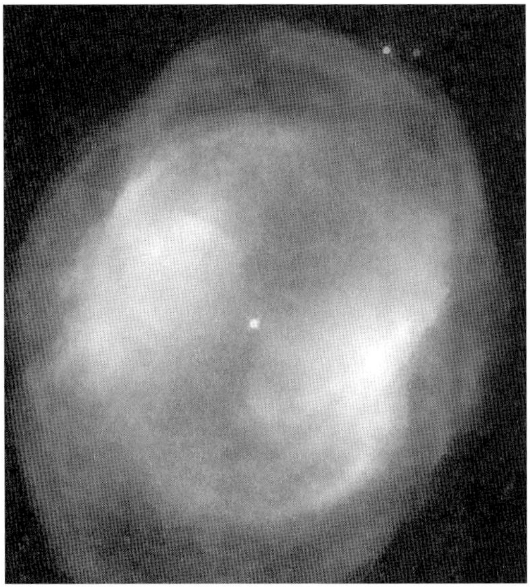

A narrow field (40 arcsec wide) Hubble Space Telescope image of the planetary nebula NGC 3195. Image: NASA/Howard Bond.

Chapter 3

How to Visually Observe the Caldwell Objects

In days gone by the majority of amateur astronomers were visual observers who trained their eyes, brains, and hands to sketch the objects they saw through the eyepiece. The planetary observers trained their eyes to see the finest details glimpsed in multiple brief moments when the atmosphere was steady, whereas the comet and nebulous object observers trained their dark adapted vision. In the era when advanced astronomy was largely the hobby of the wealthy Victorian gentleman a 6- or 8-in. (15 or 20 cm) brass tubed refractor in a substantial domed observatory was often the norm, and a custom built observing chair was always available so the observer could observe in a civilized manner.

In this era of CCD imaging, where portable telescopes are very common and observatories are often sheds on wheels, the comfortable observatory seems to have become a thing of the past, especially when a telescope can be controlled robotically in your garden or half way across the world. However, some degree of comfort is still vital for visual observing, as telescope eyepieces can invariably end up in the most inconvenient positions on the coldest and dampest of nights. For example, an f/4.5, 0.4 m Dobsonian will have a tube length similar to its focal length of 1.8 m, necessitating a ladder to reach the highest eyepiece altitudes and some neck-twisting observing positions. Observer discomfort is not good for seeing down to those very faint targets, and if you are a beginner, craning your neck to get to the eyepiece while holding a chart in one hand and a red torch in the other is no fun at all.

Alt-azimuth mounted "Go-To" Schmidt-Cassegrains offer an ergonomic alternative for the visual observer, with the big advantage that the eyepiece moves little wherever the telescope is pointed. However, if you use the supplied star diagonal you may need to refamiliarize yourself with all the star fields, as the view through the eyepiece will be mirror-flipped. This latter issue can be solved by purchasing a 45° erecting Amici prism, which will turn the view into a terrestrial one (the right way up) but will not mirror-flip the view. A thoughtful arrangement of finder telescopes on any large instrument can make locating an object in the main field a lot simpler, too. As a general rule visual observing is best when two criteria are satisfied, namely (1) the telescope should have as large an aperture as possible, and (2) the telescope should be as user-friendly as possible. Unfortunately, these two criteria often prove to be mutually exclusive.

M. Mobberley, *The Caldwell Objects and How to Observe Them*,
DOI 10.1007/978-1-4419-0326-6_3,

The ultimate observing system is the Coudé system, where extra mirrors are employed to reflect light through hollow declination and polar axes to the observer's eye. Because of the cost of making such systems, they are rarely seen in amateur hands. A compromise design, sometimes known as the Springfield system employs a Newtonian reflector with the open end deliberately counterbalanced such that the tube balance point is close to the eyepiece, so, at least in declination, the eyepiece hardly moves. Again, such systems are rare and so, in practice, an adjustable height observing chair, or even, a variable height telescope tripod or plinth are the main methods used to increase observing comfort. One particularly nice idea on Caldwell "inventor" Patrick Moore's 15-in. f/6 Newtonian is a rotating top end, which helps considerably in getting the eyepiece to a comfortable position, even if it needs a very well-engineered system to retain collimation.

Finding Deep Sky Objects

For the beginner, finding faint fuzzies can be a major hassle, not least because the night time environment is not conducive to a patient approach. It is often bone chillingly cold on the best nights, a howling gale can make things worse, and the telescope can be streaming with dew. The retina can take 30 or 40 min to fully dark adapt, and cloud, thin or thick, can terminate an observing session instantly. All these factors can sap the desire of the novice observer.

One might think that a "Go To" system is the answer, but many of the cheaper systems can miss the target and require a nightly, mind-numbingly tedious three-star alignment procedure before they can work well. The best chance of finding those really faint objects is using as big a telescope as possible, with as big a finder as possible, or even an equatorially mounted polar aligned telescope with good old-fashioned engraved setting circles. An outlay of $500 might buy you a budget 90-mm aperture "Go To" telescope or a 200-mm Dobsonian. The latter, with five times the light grasp, will give you much more success. When an object looks five times brighter you do not have to struggle as hard to see it! The R.A. and Dec. positions have been given for all the objects in the Caldwell catalog, but the reader may well prefer to locate the objects on his or her own star chart or, simply use "Go To" if he or she has a reliable system.

Setting Circles

Let us return to the subject of good old-fashioned engraved setting circles here, because their benefits are often overlooked in these days of "Go To." Without a doubt "Go To" and CCD imaging have been the biggest reasons for the massive boom in Schmidt-Cassegrain and other telescope sales since the early 1990s. The idea of not having to battle to find an object and of not having to have the vision of an owl to see an object has been a big selling point. However, "Go To" in cheaper telescopes can often be just as frustrating as starhopping and using a finder, because of the tedious star alignment procedures, the neck-cricking polar alignment telescopes, the inaccuracy of cheap plastic drives, and the unreliability of low quality encoders and gears, which are often glued together yet are asked to slew at more than 10,000 rpm!

Many inexpensive equatorially mounted telescopes have metal or plastic setting circles (see Figs. 3.1 and 3.2), which can easily be used to find faint objects simply by calibrating the circles against the right ascension and declination of a bright star and then pushing the telescope to the deep sky object's position or by calculating the offset in right ascension and declination from a bright star to that same object. It just requires a bit of R.A. and Dec. homework before you go outside. Even if your portable equatorially mounted telescope is only rough-aligned within a few degrees of the pole, this method will work well for short hops of 10 or 20° from the bright star to the object.

One word of caution here: very small setting circles often have their R.A. bands subdivided into quite crude fractions of an hour, and so you need to be sure exactly how many R.A. minutes each division stands for. In fact, in many circumstances, it can be just as easy to simply use the Dec. circle alone to set the Dec. of a relatively bright deep sky object (relative to a bright star) and, if you know the sky fairly well, just sweep about in R.A. near the part of the constellation where the faint fuzzy lives. This technique also works well for finding new bright comets in strong twilight when you have a permanently aligned telescope on a pier

Fig. 3.1. A plastic declination setting circle attached to an inexpensive Russian "TAL" telescope. Although only divided into 2° increments setting circles like this can be invaluable for finding objects without the tedious multiple star alignment menus of some "Go To" telescopes.

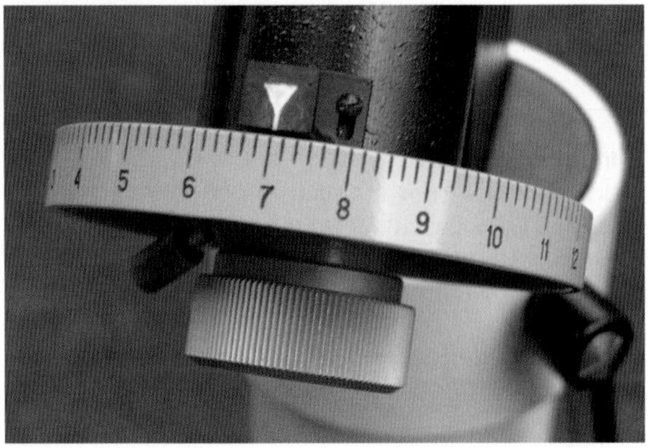

Fig. 3.2. A plastic right ascension setting circle, attached to the same inexpensive Russian "TAL" telescope. The circle is subdivided into increments of 10 min of R.A. equivalent to 2.5° of arc at the celestial equator. For a direct R.A. reading the right ascension circle needs calibrating against a star or sidereal clock, but can also be used simply to offset from a bright star. If the telescope has no sidereal motor drive in R.A., the sky will move by a degree of arc every 4 min on the celestial equator, but this should not be a big problem if the user pushes the telescope to the object immediately after calibrating on a known star.

(in other words when the declination is very accurately set). I do not wish to labor this point too strongly, but from 1980 to 2002 I used two large Newtonians (36 and 49-cm aperture) to image many faint objects, simply using setting circles, and it was a very reliable technique; far more so than using a poorly manufactured 30-cm fork-mounted "Go To" Schmidt-Cassegrain, which I also owned from the late 1990s. One final note: the "starhop" data in the column for each object is given in degrees of arc; in other words we have taken the closing up of the R.A. grid lines at high declinations into account. This was mentioned at the start of Chap. 2 but it is worth repeating. On the celestial equator an hour of R.A. corresponds to 15° of arc, whereas (for example) at 60° north or south an hour corresponds to half this angular span (as Cosine 60 = 0.5). We repeat this in case anyone with setting circles tried to recompute the R.A. offset back into minutes of R.A. without taking the cosine of the declination back into account. Remember, minutes of R.A. start to have a much smaller angle on the sky at high declinations.

The Retina

A few observers do seem to have truly exceptional night vision, but, in the main, this is very rare, and if you cannot see the faint objects that others can it is probably due to your local light pollution, impatience, a poor night, or inadequate dark adaption. Choosing the right magnification, collimating your telescope, realuminizing a Newtonian's mirrors, and, specifically, mastering the use of "averted vision" are all crucial considerations. Understanding how the human eye's detector works can help considerably, too.

The eye has two types of detector within the retina. These are called rods and cones. The rods are the low light detectors, whereas the cones allow full color high resolution eyesight. The central one degree of the retina is packed with cones that you are using to read this sentence. Your brain creates the illusion that the whole book page is sharp, but in fact you are only seeing a few letters at a time at high resolution, and in full color; it is your eye muscles that are swiftly zipping everywhere and creating the illusion. The electrochemical signals from the retina's detectors travel via cells known as ganglion cells on their way to the brain. In the high-resolution, full color retina center (the fovea), one ganglion cell interfaces to one cone. But, as you go further out and low light rods dominate, there may be a hundred rod detectors passing their electrochemical signal into just one ganglion cell. It is a case of paralleling up to improve the signal to noise ratio. Not surprisingly, with so many detectors being ganged together, resolution suffers badly. Although the foveal cones can resolve 1/60th of a degree (1 arcmin), the ganged rod system well away from the center might only resolve 20 arcmin; this is not much finer than the diameter of the Moon seen with the naked eye.

There is an optimum, ultra-sensitive, rod-packed region of the retina that is roughly 8–16° away from the eye's center; 12° is a good average value for the best part. This means that, to get the most sensitivity out of your retina, you have to look to one side of the faint astronomical object you are trying to see; at first this will seem incredibly difficult, but it will improve with practice. This 12° (or so) offset should be arranged so that you appear to place the object nearer to your nose! The reason for this strange requirement is that the eye has a blind spot where the optic nerve leaves the retina, and this blind spot is on the other side, away from the nose. (Actually, it is not! The eye's lens turns everything upside down, but we are looking at how it feels here and not how it actually is!). Roughly speaking, the eye is four astronomical magnitudes (40×) more sensitive at this crucial point than in the center. So if you can hold, say, a tenth magnitude star steady in the visual center of a 30-cm reflector's telescope field you can hold a 14th magnitude star steady on the rods 12° off center.

Dark Adaption

Of course, when you first go outside and look through the eyepiece you will probably not see anything. This is because you are not "dark adapted." When the human eye is plunged into darkness, two things happen. First, the pupil dilates (expands) to its maximum diameter. In young people, this may be 7-mm or so, but for astronomers in their eighties it may only be a few millimeters across. This is not a big problem at the telescope, as a higher magnification produces a narrower beam of light that can pass through a small pupil. A tighter beam of light will also be less affected by astigmatism in the older observer's eye.

The second development in darkness is that the amount of the chemical rhodopsin in the retina increases dramatically, by many thousand-fold. So the combined effects of rhodopsin and using averted vision amounts to a 100,000 times more sensitivity than your central vision had in a fully illuminated room before you stepped outdoors.

Dark adaption, namely waiting for the rhodopsin to do its job, cannot be rushed. You need to wait 40 min or more to feel the full effect. So, if you are planning to observe a number of faint objects, save the faintest ones till the very last!

Somewhat surprisingly, finder charts showing stars down to a mere magnitude 11 are often the best ones to use to locate the right field. After half an hour and, at very high powers, you can detect stars of magnitude 15 with a 35 cm aperture, but for that initial location of the target a wide field chart to magnitude 11 is ideal.

Spending a day outdoors in the Sun is not a good idea if you are a faint variable star observer, unless you wear wrap-around sunglasses all the time. The eye takes 24 h to recover from such an onslaught, and you will lose 0.7 or 0.8 magnitudes of your sensitivity the following night, however long you try to dark adapt. The usual 30- or 40-min dark adaption routine will not work. The same rules apply to faint fuzzies such as the Caldwell objects, as well as to faint stars.

Seeing Fainter

It might instinctively be expected that the "exposure time" of the eye was somewhere in the order of a fifth of a second, or not dissimilar to the human reaction time. Although this is true for daylight observations, in darkness things are rather different. At the faintest levels it pays to stare at an object such that the flow of photons hitting the rods is sufficiently high enough for several seconds to trigger a definite "hit" in the brain. In practice, at a typical observing site, with background light pollution, this level is usually reached when several hundred photons per second are hitting a group of rods.

This staring action should not be interpreted as an exposure time as such. The eye is not a digital device, like a CCD, but it certainly pays to stare at a faint object to see if it emerges. What the trained and dark-adapted eye sees, after studying an object patiently, is an impression not dissimilar to what a low resolution CCD image would capture with an exposure of a second or so. Of course, at these faint levels we can hardly ever make an accurate magnitude estimate of a star; all we can do is say we have seen it and maybe estimate its magnitude in relation to a slightly brighter comparison star.

Let's return to something mentioned earlier. In elderly observers, the eye's pupil does not dilate as much as the 7 mm that the young eye is capable of. This means that a loss of light results at very low magnifications. The bundle of light rays from the eyepiece will have a diameter equal to the telescope aperture divided by the magnification. If this is larger than 7 mm then light will be lost, however young you are. Older observers will find that 5 mm is a more sensible figure to adopt. In other words, with a 300-mm aperture telescope, magnifications of 60× or more are highly recommended. Much higher magnifications are typically used for studying faint stars, and the diameter of the pupil is really only a consideration at low powers when you are probably just locating the field anyway.

Many newcomers to astronomy think that to see fainter you need a lower magnification. This misconception may come from the fact that on bright objects such as the Moon, as you whack in a higher eyepiece the view just gets increasingly faint. However, different rules apply for the faintest objects. For a start, contrast becomes just as important as brightness. Only the luckiest amateur astronomers enjoy a really dark sky, and for many town dwellers the night sky has a horrible orange or yellow glow from the thousands of nearby streetlights. As you increase the telescope's magnification, the background skyglow becomes much dimmer, and point sources such as faint stars (typically spanning a few arcseconds in diameter due to atmospheric turbulence) start to cover maybe just a few more rods, convincing the brain that something is really there. Both of these factors enable the faintest stars to be seen more easily. Also, as the field of view becomes narrower, there is less chance of any really bright stars encroaching into the eyepiece field and dazzling the observer.

At this point we should mention something that is, admittedly, puzzling. Most knowledgeable amateur astronomers will tell you that a telescope's focal ratio has no bearing on how bright the background sky appears in a light polluted area. The sky background will simply get dimmer and blacker as you whack up the magnification. If an f/6 and an f/4 Newtonian of the same aperture are used at the same magnification, the sky background brightness should look the same. Obviously to match the magnifications you would have to use two different eyepieces, such as a 4-mm eyepiece with the f/4 instrument, and a 6-mm eyepiece with the f/6 instrument. However, if you ask really experienced visual observers who live in highly light-polluted areas what their opinion is they will tell you that the slower f-ratio telescope (f/6 in this example) always makes the sky darker, even at the same magnification. This has to be an issue more related to scattered light in the tube than a straightforward optical issue. The f-ratio of a telescope should have no bearing on the background sky brightness for a specific magnification. The same number of photons will hit the same number of rods and cones when aperture and magnification are identical unless there is something different in the design of the two telescopes or the two eyepieces. Nevertheless, this "slower is darker" rule does apply, assured by an observer who has made more than 200,000 magnitude estimates down to (almost) magnitude 17, so you should take it seriously. The performance of eyepieces on telescopes with fast optics may have a bearing on this issue, too.

So, just how faint can the human eye reach? Well, for typical observers in good conditions the standard stellar formula that is most often used is $2 + 5\ \log_{10}D$, where D is the telescope diameter in millimeters. However, nothing in life is simple, and experienced observers can see well below the magnitudes predicted by the standard formula. Indeed, some can see stars to almost magnitude 8 at the zenith simply using the naked eye at a truly dark site. Unfortunately, human beings are impatient, intolerant of things that cannot be stated precisely in black and white. People like a straight answer to a straight question and not "well, it all depends on" types of answers. The plain fact is that every observer has a different degree of experience, is faced with a different amount of light pollution, has a different set of optics, and has a different pair of eyes. So the answer to "How faint can I see?" has to be "Go outside at night and find out"! With extended deep sky objects such as those in the Caldwell list things become even more vague, as we are talking about magnitudes per square arcsecond, not just point-source magnitudes. A crystal clear dark sky and some nebula filters become even more important.

One authority on human night vision, Roger N. Clark, has pointed out (*Sky & Telescope* April 1994, pp. 106–108) that a study conducted by H.R. Blackwell in 1946 likened the eye's detection ability, at its limit, to a probability curve. To quote specific examples, Clark proposed that with a 400-mm aperture telescope an observer working at the limit, in excellent conditions, could just glimpse a magnitude 15.7 star 98% of the time, a magnitude16.7 star 50% of the time, and a magnitude17.7 star 10% of the time. The 50% probability level corresponds to the formula $3.7 + 5\ \log_{10}D$ and the row labeled "Clark" shown in the table below. In fact, Clark's 50% probability formula corresponds very well to the sorts of magnitudes reached by the world's most experienced visual CV observer, Gary Poyner, who has glimpsed stars well into the high magnitude 16s with a 35-cm Schmidt-Cassegrain and with 40 or 45-cm Dobsonians.

Table 3.1. Various visual stellar magnitude limits for telescope apertures in millimeters, predicted by formulae proposed by astronomers in recent years

Formula	100 mm	150 mm	200 mm	250 mm	300 mm	350 mm	400 mm	450 mm	500 mm
Standard	12.0	12.9	13.5	14.0	14.4	14.7	15.0	15.3	15.5
North	13.3	14.1	14.6	15.1	15.4	15.7	15.9	16.2	16.4
Schaefer	13.4	14.3	14.9	15.4	15.8	16.1	16.4	16.7	16.9
Clark	13.7	14.6	15.2	15.7	16.1	16.4	16.7	17.0	17.2

The standard formula has always been 2.0 + 5 log10*D* (where *D* is the telescope diameter in millimeters) despite the fact that many observers can see fainter. Gerald North's formula (4.5 + 4.4 log10*D*) was first suggested by him in the *Journal of the British Astronomical Association* in 1997 (J. Brit. Astron. Assoc., 107(2), 82 (1997)). It fits Schaefer's results rather better for a wide range of apertures than do other formulae and, although more optimistic than the standard model, becomes slightly more pessimistic as very large apertures are used. The third row uses Bradley Schaefer's own prediction (3.4 + 5 log10*D*), while Clark's (3.7 + 5 log10*D*) is more optimistic and ties in with his 50% probability criteria

The other rows in Table 3.1 correspond to a formula proposed by Bradley E. Schaefer (NASA-Goddard Space Flight Center) in 1989 after an extensive survey, and a formula proposed by the author and amateur astronomer Gerald North, after studying Schaefer's report. Of course, there are always the really freaky results that amateurs tend to remember; for example, Stephen O'Meara's detection of Comet Halley, visually, at magnitude19.6, with a 60-cm telescope at Mauna Kea in Hawaii in January 1985, while breathing oxygen! He also identified a magnitude 20.4 field star at the same time. Although this sounds impossible he had spent "1–2 h" staring at the field waiting for a definite "glimpse." The feat was verified by O'Meara describing field stars with no prior knowledge of their position. This sort of detection feat would correspond to about 1 or 2% on the Clark "probability level".

In practice, experienced observers can often reach two stellar magnitudes fainter than the standard formula for stars at high altitude and on the clearest nights, even from urban locations. In fact, rods are so sensitive that they can actually detect single photons. In 1942 Selig Hecht proposed this because individual light flashes so dim that only 1% of rods likely to absorb a photon were detectable by observers in experiments. Experiments by Schneeweis and Schnapf in 1995, using monkey rods, confirmed that single photons could trigger a response. The arrival of a few photons per second at the eye is the sort of rate that an observer using a 400-mm telescope might receive from a 20th magnitude star, making O'Meara's famous Halley observation seem more feasible! Fortunately, the human retina has no electronic readout noise and does not need to be cooled.

Truly, the human eye is a quite remarkable detector. Even in the twenty-first century era of CCDs and webcams, its versatility is extraordinary. It can cope with illumination levels from bright sunlight to starlight, spanning 100 million-fold in intensity and even survey almost the whole night sky for meteors, following them instantly as they whiz across the sky. It is amazing what hundreds of millions of years of evolution can produce!

Eyepieces

Eyepieces have already been mentioned in connection with magnification and the benefits of low and high powers. Having a range of eyepieces giving magnifications as low as, say, 3× per inch (1.2× per centimeter) up to 50× per inch (20× per centimeter)

enables you to use the lowest power as a sort of finder and the highest power to seek out the lowest surface brightness objects, even before you start screwing those deep sky filters into the eyepiece barrel. As most amateur astronomers will know, dividing the telescope focal length by the eyepiece focal length gives you the magnification; thus a 2,000-mm focal length Schmidt-Cassegrain (200-mm aperture, f/10) with a 20-mm eyepiece gives 100× (2000/20 = 100).

There is a truly bewildering array of eyepieces available to the amateur astronomer these days, and it can be quite confusing working out what you need. The term "apparent field" tells you what the field presented to your eye actually feels like. Thus, an old fashioned 30° apparent field eyepiece will feel very much like you are looking through a narrow tube, whereas an 80° field eyepiece will make you feel like you are floating in space. Divide the apparent field by the magnification and you get the real field covered on the sky; so, with an 80° "apparent field" eyepiece, a magnification of 160× will give you half a degree (80/160 = 0.5), just enough to fit the whole Moon in.

Needless to say, the wider field eyepieces tend to be the most expensive. There seems to be a view that an eyepiece with a 2-in. (50.8 mm) barrel diameter is the ultimate you can aspire to. However, at high magnifications and with short focal length eyepieces a 2-in. barrel is not needed, as the smaller 1.25-in. (31.7 mm) barrel eyepieces are wide enough to accept the image. This is reflected in manufacturer's ranges. If we take the TeleVue range as an example, their modest apparent field 50° Plössls only requires a 2-in. barrel for focal lengths longer than roughly 31 mm, whereas their 65° wide field eyepieces require a big barrel above 24 mm, and the 84° Naglers need a big barrel above 18 mm. Their extraordinary 100° Ethos eyepiece requires a big barrel above 15 mm.

Now, while a 100° field is an amazing technological achievement, delivering a true spacewalk feeling, in practice there are other, far more important considerations when you are outside on a cold night. For one thing, some of these huge 50 mm barrel eyepieces are absurdly heavy, weighing almost 2 lb (900 g). In addition, some of them have flat tops that are so ludicrously wide that your nose has nowhere to fit and you have to tilt your head to look through them! Also, there is the technical parameter known as "eye relief," namely how many millimeters can you move your head back before your eye is outside the optimum position. Those "spacewalk" fields are only achievable when your eye is positioned at precisely the right point.

All things considered most amateurs tend to prefer a set of lighter, more manageable 31.7 mm barrel diameter eyepieces with, say, a 60° apparent field, and good eye relief. They also prefer an eyepiece with a tapering top end (that avoids the nose) and a soft rubber eye cup that excludes stray light (such as direct glare from a street lamp). For spectacle users, the eye relief issue becomes critical, as their glasses prevent them from getting too close to the eyepiece's eye lens anyway. You might think that removing spectacles and simply refocusing might dispense with the need for glasses at the eyepiece, but if you suffer from astigmatism refocusing is not a solution. However, even here TeleVue has an answer. Its Dioptrx lenses can be ordered based on your eye-test astigmatism prescription and fit on top of many of their top quality eyepieces.

So how does all this have a bearing on observing the Caldwell objects? Well, in practice, to enjoy any long-term visual study of a celestial object requires comfort, and things can be bad enough out there anyway, in the freezing dew-sodden conditions, so the last thing anyone wants to have to do is juggle with 2-lb eyepieces and have to position his or her head within a few millimeters. You should always choose a user-friendly eyepiece that fits snugly into your eye socket, excludes stray light, and gives plenty of eye relief, well above pure field of view and ornamental value. A decent set of 31.7 mm barrel eyepieces with long eye relief will be a joy to use compared with some of the monsters you can buy. They will often be easier on the wallet, too. The TeleVue Radian eyepiece ranges are especially good in this regard, even at short eyepiece focal lengths, as are the Vixen Lanthanum range, and all of the 31.7 mm TeleVue models are light and beautifully made. One other factor to bear in mind is that rather than buying a very short focal length high power eyepiece, a Barlow or Powermate lens is often a much better investment, as it can turn your medium focal length eyepiece into a high power eyepiece with much better eye relief.

Charts and Filters

Observing any form of faint fuzzy object in the night sky depends on a number of factors, not least your experience, dark adaption, suitable finder charts, dark and crystal-clear skies, sufficient aperture, and appropriate filters. Of course, with many amateurs now owning "Go-To" telescopes the finder charts might be considered superfluous; however, when tracking down an elusive object on the limit of vision, identifying the stellar background field first does give you an enormous amount of confidence that you are looking at the right patch of sky.

For the beginner to astronomy it might, at first, be hard to grasp how a nebular filter that lets through *less* light can possibly be an advantage when you are straining to see such a faint object anyway. Surely, blocking light is not the answer? In fact, it *is* the answer because emission nebulae emit light at specific wavelengths, which you want to see, but light pollution also emits light at specific wavelengths that you do NOT want to see. Ideally you want the background sky brightness to be black, to give the maximum contrast between nebula and sky. But filters can only do a certain amount. From a truly dark country site you may well be able to spot the most popular nebulae with no filtration at all. The Ring Nebula (M57), for example, is a hard object to miss through any modest telescope, even from an average observing site. For open clusters, globular clusters, and galaxies narrowband filters will not assist, as these targets emit across the visual spectrum. However, a standard visual "deep sky" filter that blocks the streetlight emissions may still greatly assist those with a serious light pollution problem.

The best nights for observing any faint objects are ones that are free of atmospheric haze. In northern areas, this usually occurs after a cold front has passed through the country (from north to south) and quite often after it has rained, and when there is a northerly air flow. Such conditions are often poor for planetary observers but perfect for deep sky work. It goes without saying that the sky should be Moon-free, too! Dark-adapting for at least half an hour (as mentioned earlier) is recommended for any faint object detection, as is experimenting with different power eyepieces. Using a software planetarium package to print your own star charts to match the field of view and magnitude limit of your favorite telescope/eyepiece combination is strongly advised.

Filter Types

A variety of visual filters are available for the deep sky observer from companies such as Meade, Celestron, Orion, and Lumicon. Essentially these boil down to four specific types, namely: deep sky/broadband nebular filters, ultra-high contrast (UHC)/narrowband nebular filters, Oxygen III (OIII) filters, and hydrogen beta (H-beta) filters. The aim of all these filters is to block sodium, mercury, and airglow emissions in the orange, yellow, and blue parts of the visible spectrum but allow through the astronomical emission lines of Oxygen III (green/cyan at 496 and 501 nm), Hydrogen beta (cyan 486 nm), and Hydrogen alpha (deep red 656.3 nm). In essence the ultra-high contrast/narrowband filters are simply a higher contrast version of the deep sky/broadband filters, and the Oxygen III filters are so high contrast that they pinch much of the astronomical Hydrogen beta line off, too. The highly specialized Hydrogen beta filters just let through that specific band, and they are only useful on a few objects at most, but, in particular, Orion's Horsehead Nebula.

In practice, the visual emission nebula, supernova remnant, and planetary nebula hunter will benefit most from a UHC/narrowband or OIII filter (see Figs. 3.3 and 3.4). Frankly, there is little to choose between them for many objects such as planetary nebulae. The OIII gives a slightly dimmer but slightly higher contrast view, which can be useful in cluttered star fields, but both dramatically enhance the unfiltered view. A standard deep sky/broadband nebular filter will increase the contrast compared with the unfiltered view and allow normal objects such as stars or galaxies to be seen, too, but the UHC and OIII filters will really make those nebulae stand out. Having both UHC and OIII filters enables you to pick and choose the one that works best for each nebula under scrutiny and for your specific light pollution problem. Sometimes the UHC will work best, and sometimes the higher contrast OIII will outperform the UHC. It is very fortunate for astronomers

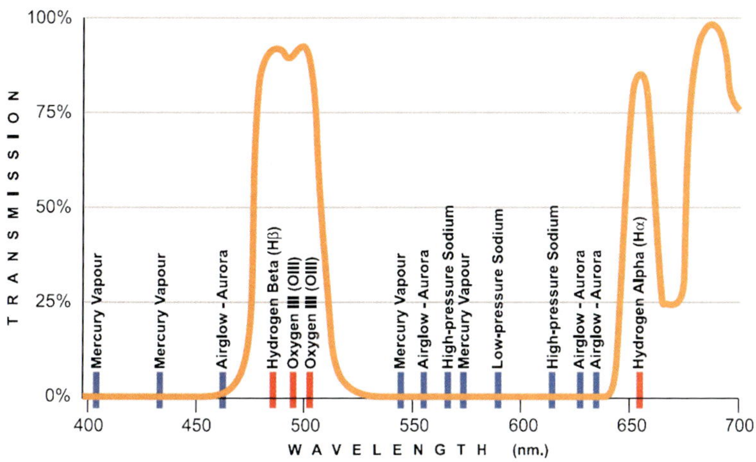

Fig. 3.3. An ultra high contrast (UHC) narrowband filter has the transmission characteristics shown in this graph. The astronomical wavebands are let through, but the sodium streetlight, mercury streetlight, and natural skyglow wavebands are filtered out. Diagram provided by Astronomy Now/ Greg Smye-Rumsby.

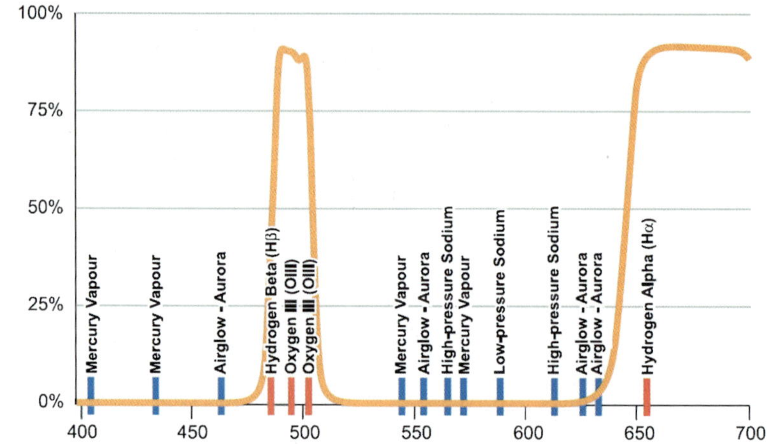

Fig. 3.4. An oxygen III [O III] filter gives slightly more contrast than a simple narrowband/UHC filter, shown in the previous figure. The Oxygen III and hydrogen-alpha wavebands are let through, but the sodium, mercury, and natural skyglow wavebands are filtered out. Diagram provided by Astronomy Now/Greg Smye-Rumsby.

that the emission lines in astronomy are in a different part of the spectrum to the mercury and sodium streetlights!

Sketching

Despite the advent of the CCD era, a hardy band of enthusiasts still sketch deep sky objects (see Fig. 3.5). Sketching is, of course, the only way to record exactly what your eye and brain observed at the telescope. It goes without saying that you do not *have* to sketch faint fuzzies if you do not want to! As a minimum, though, most visual observers will want to have a record of when and where they observed an object and what telescope and eyepieces they used. If a sketch is just too much hassle on any given night, or if cloud intervenes, then a few simple notes in an observer's log will at least preserve the observing memory.

Sometimes CCD imagers are criticized by a small minority of visual observers for not being "real observers"! This criticism is unjust. Most visual observers are not good artists, and even in the heyday of the visual observer only a handful of skilled observers could really observe and sketch with any degree of positional accuracy. It is this ambiguity in the recording process that largely turns visual observers (as this author once was) into photographers or digital imagers. CCD users are people who want to push their backyard telescopes to the limit, with the cleverest technology available, or people who want to contribute unambiguous science. Admittedly the best CCD equipment is expensive and therefore out of the range of many amateurs, but it would obviously be wrong for envy to bias opinion on this subject, especially when entry-level DSLRs are so affordable. Nevertheless, carrying out your hobby with a minimum of expenditure can bring a certain satisfaction of its own, especially if your Dobsonian is homemade, too.

There is also a myth that CCD imagers are not visual observers! This is absurd. Just about all digital imagers get a thrill from visual observing, but the frustration

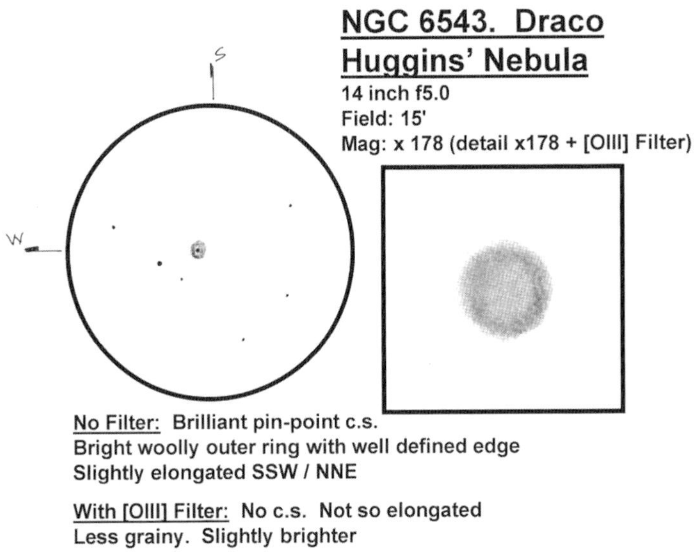

Fig. 3.5. A sketch of tiny NGC 6543 (Caldwell 6) by the experienced UK observer Stewart Moore using a 14-in. (35.6 cm) Newtonian at ×178. The central star was seen clearly without an OIII filter but disappeared when it was employed, although the nebula itself was enhanced using the filter. As with many northern hemisphere visual sketches the field is shown with south at the top.

of trying to sketch the object in a cold, dark, and damp environment and the quest to go deeper in magnitude fuels these former visual observers on to imaging. A compromise digital sketch can be created if you are hopeless at drawing; see Fig. 3.6 for one example. There are times, though, when the tangle of wires, software problems, and complexity of the whole imaging lark can irritate even the hard-core CCD user. It is then time to give the technology a rest, get the Dobsonian out, and just enjoy a night under the stars, observing with real photons hitting the retina. Nothing is more relaxing, even if you are heavily into CCD imaging.

Some observers (with civilized observatory domes and a comfortable, seated viewing position) make regular sketches at the eyepiece; many others make very rough notes outdoors, accompanied by an audio recording, and then do the bulk of the sketch indoors, in the warmth. One of the major hassles with making any kind of sketch is simply getting comfortable at the eyepiece! Despite the boom in astro-merchandise in recent years, there is still no one making a really comfortable fully adjustable observing chair!

The Sketching Process

Whether you make a drawing outside, or inside from copious notes, various sketching techniques can be employed to make a nebulous object look as fuzzy and indistinct as it did through the eyepiece. A soft, 4B pencil is a good choice, and a nebulous look can be imparted by rubbing the sketch with a finger, an artist's stub pencil, or a cotton bud. Pointed sticks known as tortillons or blending stumps can also be purchased if you get serious about drawing fuzzy objects. You can rub these devices in a load of black graphite scribbled at the edge of your sketch pad with a

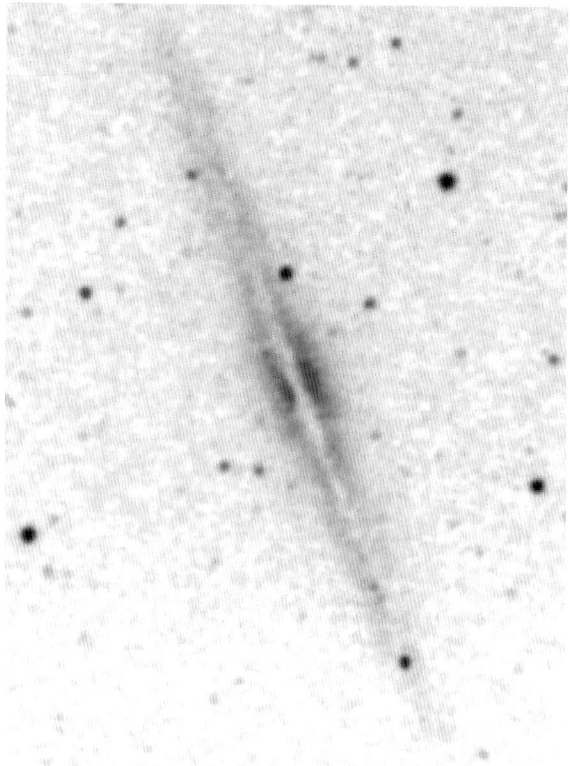

Fig. 3.6. An attempt by the author to simulate the visual appearance of the edge-on galaxy NGC 891 (Caldwell 23) through the eyepiece of his 35 cm Celestron 14 at ×200. Positional accuracy from digital images with the same instrument.

soft pencil and then use them as a poor man's airbrush. If a sketch on white paper is scanned and then made negative, the effect can be very realistic. Harder pencils have their uses, though, especially for drawing faint stars that are just glimpsed as points. Experimentation with pencils and paper can be most useful, and artist grade papers will yield the best results. For damp nights you may want to keep a secret stock of waterproof paper, too. This does really exist! Obviously some type of clipboard or hard platform will be required to fix the paper on a solid drawing surface. The choice of eraser is important, too, as some colored versions leave an unwelcome stain on your pristine sketch; a white one is best. It can be interesting to make a sketch and then compare it to a digital image to understand just what the eye thinks it has seen; see Fig. 3.7 for an example.

Although some observers have used white pencils on black paper for sketching deep sky objects most use black pencils with white paper. Thus, stars are black circles, with the brightest one being more than a few millimeters across and bright nebular regions appear as intense dark shading. Nearby bright security lights (sometimes called Rottweiller lights in the UK) can be intensely distracting when trying to see faint objects at the eyepiece. Some observers have been known to drape a black cloth over their heads to exclude such lights, although the rubber eyecups on the more expensive eyepieces can be quite effective. The problem with the black cloth approach is that most telescope eyepieces are much colder than the

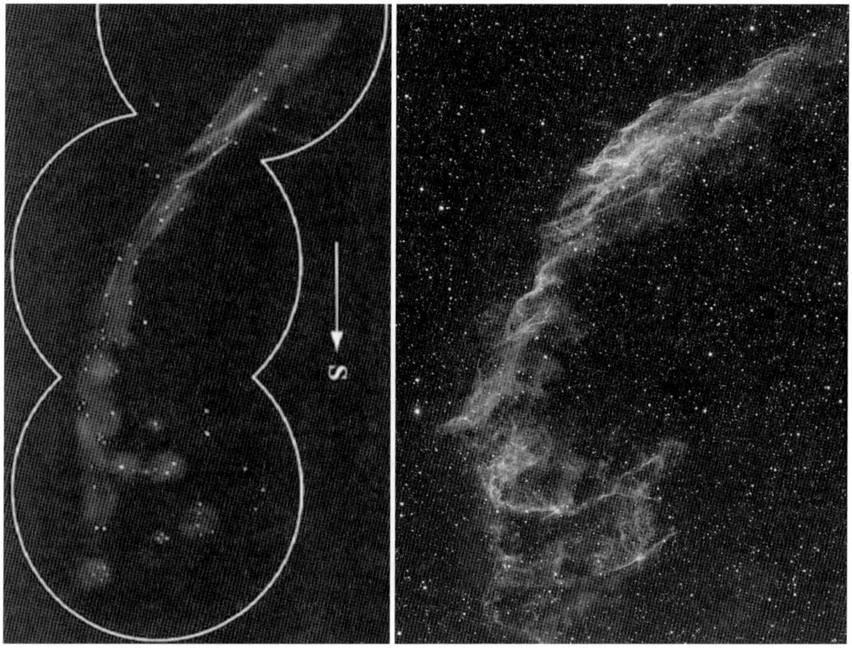

Fig. 3.7. A sketch by the experienced visual observer Stewart Moore (left) of the easternmost part of the Veil nebula (NGC 6992 upper and NGC 6995 lower) also known as Caldwell 33, compared with a digital image (right) by Ian Sharp shown earlier. Stewart used a 14-in. f/5.0 Newtonian at ×100 with a UHC filter and needed three half degree eyepiece fields to fit the Veil in. The naked eye limiting magnification was 5.5. Stewart's description reads: "UHC filter – superb. Too much detail to draw. Parts look three-dimensional; Sinuous twisting bands of varying intensity. IC 1340 suspected. OIII filter – Disappointing. Great loss of detail compared with UHC filter." As noted in the section on Caldwell 33, Ian Sharp's ATIK 314L CCD image mosaic was taken with an 80 mm semiapochromatic Orion ED80 refractor (600 mm focal length, f/7.5). Each half of the mosaic consists of 18 ten-min exposures.

observer's breath and so will mist up rapidly if the observer's head and the eyepiece are enclosed in the same space.

Of course, the observer needs to be able to see charts and sketch the object in question, and so some sort of very low level illumination is required. The human eye is less dazzled by red light, and so many observers use red film, plastic, or cellophane across a standard torch to provide a level of lighting that will not wreck the dark adapted night vision. In some places, red bicycle rear lamps are quite popular. If you can devise a dimming circuit for your red light, with the whole unit clippable to your clipboard, this will be of huge advantage. There is a fine line between too faint and too bright, and this line will vary as your eyes dark adapt. Variable star observers who have learned the tricks of dark adaption often make good deep sky observers, too; see Fig. 3.8 for one example of a sketch by a hardcore variable star man.

To be of value a sketch should contain a minimum amount of information to enable it to qualify as a scientific observation. The date and time of the observation are essential, along with the telescope type, aperture, and f/ratio, plus the magnification used and the field of view covered by the sketch. Quoting the double date (such as September 20/21) will avoid any later confusion that can often arise when the time of observation was close to midnight but your brain was saying it was still

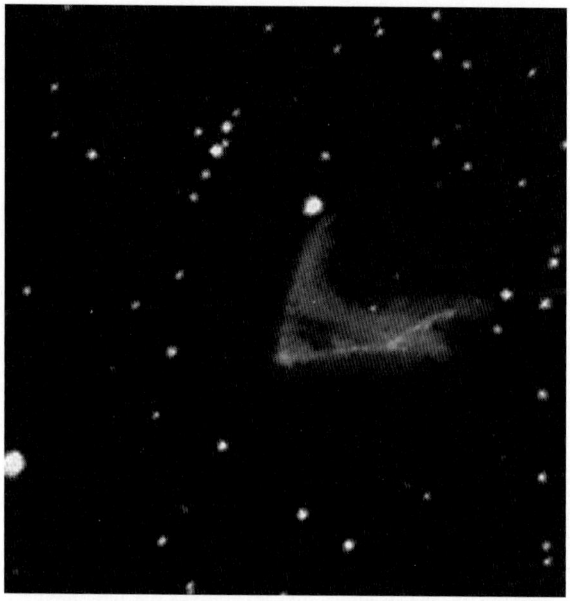

Fig. 3.8. A sketch by the eagle-eyed UK variable star observer Gary Poyner of Hubble's variable nebula NGC2261 (Caldwell 46). Sketch made on March 1st 2003 at 22 h UT with an 18-in. (46 cm) Dobsonian.

yesterday's date, despite having gone past midnight. U.T. (Universal Time) should always be quoted for the time, and it is a good idea to permanently set an observing clock to U.T.

Vital Data for the Log

The field of view of an eyepiece attached to a telescope can be calculated with ease by first calculating the magnification (telescope focal length divided by eyepiece focal length) and then dividing the eyepiece's advertised apparent field (typically 50–80°) by the magnification. So a 2,000-mm focal length telescope with a 20-mm eyepiece will have a magnification of 2,000/20 or 100×, and, with a 60° apparent field eyepiece the field will be 60/100 = 0.6°, or 36 arcmin.

If you do not trust the eyepiece or telescope data, then finding a star on the celestial meridian (zero declination) and timing its transit from one side of the eyepiece field to the other, with the drive turned off, will give you the eyepiece field, too. So, for example, if a star at zero declination takes exactly 4 min of time to cross the field, the field is exactly 1°, as the sky moves a quarter of a degree per minute. You can do this at any declination as long as you factor in the cosine of the object's declination. For example, at 60° Dec. (Cos 60 = 0.5), if a star takes 4 min to drift through the field with the drive off, the field of view is exactly half a degree because the sky drifts past your eyepiece half as fast. If you prefer to express the eyepiece's real field of view (RFV) in arcminutes then just multiply the drift time by 15, so a 4-min drift = 60 arcmin on the celestial equator. The direction of north, south, east, or west should also be recorded, and this can be complicated by the use of a 90° star diagonal commonly used with Schmidt-Cassegrains and refractors. Some observers just indicate the direction the object drifts, namely to the west, with the drive switched off, to avoid any confusion. But, if a star diagonal is being used this should be indicated.

If you have a portable observing system then the observing site should be recorded. Perhaps the most important entry is simply one noting the sky conditions at the time of the observation. Years later you may recall that, in the past, you had a much poorer or better view of the same deep sky object and be mystified why that was. If your logbook tells you that the sky transparency was poor, or that the first quarter moon was in the sky you will forever know why that observation was rather poor. Many observers quote the limiting magnitude in their notes, in other words, the faintest naked-eye stars they could see near to the object in question. If it was, say, 5.5 from a typical rural site, then all was well, but if it was only 4.0 it would have been a grotty, hazy night. For small objects, such as planetary nebulae, the atmospheric "seeing" will affect what you can resolve.

Admittedly, poor seeing is more usually associated with lunar and planetary observing, where you are trying to eek out the detail visible at the limit of your equipment. But when trying to see, for example, a faint central star in a planetary nebula, it can be critical, too. Many observers use the Antoniadi scale of seeing, which is defined from I to V with I denoting perfect conditions at high power, without a quiver in the eyepiece (very, very rare indeed) and V indicating that stars are just a fuzz ball, being as large as 10 arcsec across in some cases. Good atmospheric seeing and sky transparency are often mutually exclusive events, as the most stable conditions quite often occur when the sky is misty, but, all sorts of combinations are possible and should be recorded. It goes without saying that any filters used to enhance the view should be noted, too, along with their enhancing effect. All modern eyepieces feature threaded barrels into which deep sky filters can be screwed. An alternative approach is a filter wheel or a manually operated filter slider.

Of course, different types of object require different observational and sketching techniques, and the experienced visual observer will be asking different questions at the eyepiece before committing pencil to paper. For galaxies the texture of the object is important. Does it look mottled after sufficient dark adaption? Do the galaxy edges fade into the background gradually or end abruptly? What foreground stars are visible close to the galaxy (this is crucial for supernova hunting, and only the brighter stars visible in a CCD image will be glimpsed by the retina)? If the galaxy is not edge-on, is there evidence of a spiral or barred-spiral structure? If it is edge-on, can the central bulge be seen visually or even a dark central dust lane?

In contrast, for open star clusters there may well be a feeling (illusory or otherwise) that there are chains of stars running through the field. Conversely, the observer may feel that there are star voids in other parts of the field. These effects are rarely seen as well in images because you are seeing what your logarithmic response retina and human brain feel is there, right up to the visual limit, rather than the cold digital response of a CCD. Representing, on paper, a wide range of star brightnesses, as they appear to the eye, is a major challenge to the artist, and it is easy to forget the brilliance of the brightest stars when the observer returns indoors. CCD images, while going much deeper, do not convey the visual punch of the brightest stars as well as they actually appear, but short CCD exposures can help a useless sketcher compile a digital impression of what they saw; see Fig. 3.9

Fig. 3.9. An attempt by the author to simulate the visual appearance of the globular cluster NGC 6934 (Caldwell 47) through the eyepiece of his 35 cm Celestron 14 at ×200. Positional accuracy from digital images with the same instrument.

for an example. Of course, to the eye, observing is a realtime experience, with bright stars twinkling and scintillating with the seeing. It is horrendously difficult to translate this feeling onto paper.

Then there are globular clusters. Again, in large telescopes the eye may spot distinctive chains or patterns of stars that simply do not leap out of a digital image. The human eye and brain combination is very good at seeing patterns, especially human faces, in chains of dots seen through the telescope. Globular clusters vary in how compressed their centers are, and while CCD images can tend to saturate their cores the visual observer tends to get a much better feel regarding just how the star density increases at the center, equivalent to a snapshot CCD image of a second or so. With large apertures, the brighter stars can seem to sparkle because they are being seen with direct vision, not just suspected with averted vision, as for the faintest stars – all very tricky to represent on paper.

Such is the often uncomfortable observing position that a deep sky observer finds himself or herself in that making any kind of meaningful sketch at the time is often impossible. Let us be realistic here; you can be up a ladder or with your neck twisted at a painful angle. You may be wearing arctic thermal clothing and dew or ice may be forming on the eyepiece. In addition you are often tired and are trying to hold onto a ladder with one hand, a torch with the other, and a pencil and notepad with the third hand that you don't have! Hardly surprising, then, that many observers just make rough notes, a very rough sketch, and use a digital recorder to make other notes verbally. Of course, in this digital era, it is all too tempting to use a CCD image from the web to faithfully record the salient features as a template on which to base your sketch.

Most regular observers will take pre-prepared sheets outside with a circle already drawn on them representing the eyepiece field of view (see Fig. 3.10, for example). Do not assume that you need to sketch everything at the eyepiece.

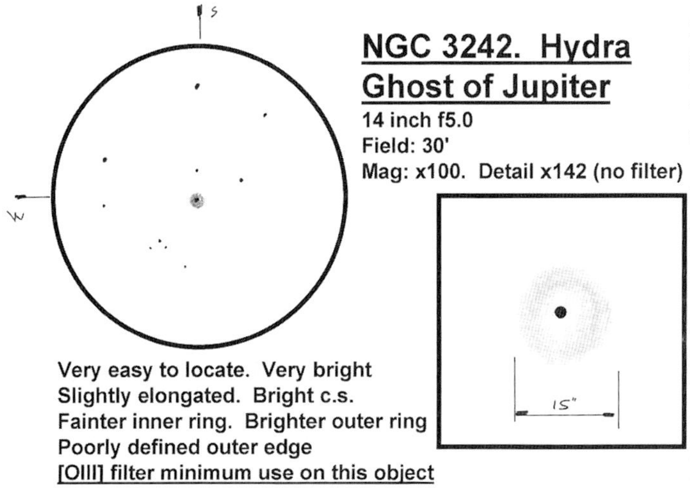

Fig. 3.10. Another sketch by the UK observer Stewart Moore with his 14-in. f/5.0 Newtonian. This is of another planetary nebula, NGC 3242, or Caldwell 59, the "Ghost of Jupiter."

Many mottled shapes seen in nebulae are simply impossible to draw outside, and so a rough sketch, accompanied by copious notes or an audio recording, will suffice. For the best positional accuracy visualizing a clock-face, or a system of concentric rings within the predrawn circle, works well. Indeed, if you are planning to completely redraw the sketch you could sketch stars and features within a pre-prepared template on your paper. You might also consider drawing brightness contour lines (isophotes) to define the radial intensity of a nebula; once indoors these contour lines can be used as a reference to create a smoother look to the nebula as it fades into the background sky.

Chapter 4

How to Observe the Caldwell Objects Digitally

Imaging deep sky objects is a subject in itself. Indeed, entire books have been written on the art and science of digital astrophotography. However, most of these books contain far more information than the Caldwell imager will ever need, and, in practice, mastering your own equipment and software (not someone else's) is the key to success.

Of course, everyone has different equipment, and the biggest bugbears are always those that involve hardware and software incompatibilities. Fortunately, the Internet has spawned a whole host of technical user groups that you can tap into to obtain a wealth of hard-earned knowledge from other amateurs who have learned the hard way and are keen to help beginners using the same equipment. To tap into this enormous knowledge base you first need to visit: http://groups.yahoo.com/search?query=astronomy, which will give you a massive list of technical astronomy user group sites that will almost certainly include some relevant to your equipment. For example, one of the most popular CCD manufacturers, SBIG, has a user group at: http://tech.groups.yahoo.com/group/SBIG/

The UK CCD camera manufacturer Starlight Xpress have a similar group at: http://tech.groups.yahoo.com/group/starlightxpress/

Consulting with such groups should enable you to cope with those infuriating USB, Windows, and hardware hassles that form the first hurdles to taking great images.

Monochrome, Filtered Color, or One-Shot Color?

The first decision the astrophotographer will make is whether to take monochrome or color deep sky images and, if color is the aim, whether a one-shot color CCD camera or DSLR will be used, or if advanced filtered work is anticipated. (DSLR stands for digital single lens reflex: a digital camera with a removable lens.) For many beginners a simple entry-level DSLR will be their starting point into astronomy.

M. Mobberley, *The Caldwell Objects and How to Observe Them*,
DOI 10.1007/978-1-4419-0326-6_4, © Springer Science+Business Media, LLC 2009

With a T-adaptor to interface the camera's bayonet mount to a 2-in. or 1.25-in. adaptor (50.8 or 31.7 mm) you are in business, and the standard approach will be to take 30 or 60-s exposures (depending how good your telescope drive is) and stack the resulting JPEG images up in *Registax* or another astronomical package. It should be kept in mind that one long exposure will always go deeper than the equivalent duration in multiple short exposures, but the time you can expose for will depend on your telescope drive and the focal length you are working at.

Key Factors

There are a number of key factors that need to be in place before a truly great astronomical image can be produced, rather than the run-of-the-mill routine astro-images that are just "snapshots" of the sky. The darkness of the night sky from your location is a major factor but, unfortunately, not one over which you have much control. However, with advanced filtered work many of the effects of mild light pollution can be compensated for on those crystal clear nights when a cold front clears the light reflecting dust out of the atmosphere. Critical focusing is also crucial to any good astronomical image; you may think we are stating the obvious here, but it is surprising how often focusing is neglected. The quality of the imaging camera is critical too, as is the temperature the camera is running at. With DSLRs not having any cooling option they can be running at 20°C on a mild night, whereas a cooled astrocamera might be running at –20°C, giving at least a 50× improvement in electronic thermal noise.

As mentioned earlier, the quality of the telescope drive is a major factor, especially in long focal length work. Autoguiders, which detect the star drifting on the CCD pixels and adjust the telescope mount to compensate, are often used to correct drive errors, but they do take time to set up on a suitable guide star, even when they are built into the main imaging camera head. The most advanced amateurs even employ a form of adaptive optics whereby a small, lightweight, rapidly tilting mirror reacts in a split-second to a drifting star (far quicker than a heavy telescope mount) and so virtually eliminates tracking errors.

Atmospheric seeing is another factor in image quality. On the best nights the bulk of a faint star's light may spread over only a couple of arcseconds, but on the worst ones it may spread over 10 arcsec. Unfortunately, those crystal clear nights when transparency is at its best and light pollution is minimized often coincide with very jittery seeing – sometimes you just cannot win! Because of this seeing issue many experienced amateurs arrange their equipment to give an optimum "sampling" ratio of around 1–2 arcsec per pixel but no finer, thereby resolving details of a few arcseconds in size, resulting in the maximum resolution the atmosphere will permit, but not losing magnitude penetration by working at uneccesarily long focal lengths. Of course, achieving this optimum sampling requires a careful choice of CCD camera such that the pixels are just the right size for your focal length, with or without some form of optical telecompressor. There is a handy formula that enables you to calculate the sampling ratio of your system, namely:

Sampling in arcseconds per pixel = 206 × pixel size/focal length

This formula uses pixel sizes in microns (1 μm = 1/1,000 mm) and focal lengths in mm. For example, for a 356 mm aperture Celestron 14 working at *f*/7.7, and a focal length of $356 \times 7.7 = 2741$ mm, using a camera with big 20 micron pixels the formula gives: $206 \times 20/2{,}741 = 1.5$ arcsec/pixel.

The final major factor in the quality of any astronomical image is the application of advanced color filtering and software techniques that will be discussed later.

The Problem with Color

Monochrome deep sky images, taken with an unfiltered detector, are, pretty much, hassle free. However, the same cannot be said for color images. Objects in the night sky are incredibly faint, and, in any form of color imaging, filters are placed in the light path to extract the color information. In commercial color CCD or CMOS sensors, as used in DSLRs, a "Bayer matrix" typically composed of a 2×2 array of filtered pixels is used to extract the color information. With this system every pixel in the chip is behind a miniature, microscopic filter, such that two diagonally opposed pixels in each 2×2 grid are beneath green filters and the remaining two pixels sit behind a red and a blue filter. In addition, the whole sensor sits beneath an IR-UV blocking filter, which stops light from the extreme infrared and ultraviolet getting to the CCD (thus mimicking the eye's sensitivity).

Some amateurs use DSLRs where this blocking filter has been removed to enhance red nebulosity; however, the colors of daytime objects are then seriously skewed, and the focus point through the optical viewfinder may not match that on the chip. Essentially, the use of a filtered DSLR results in a loss of around 1.5 magnitudes compared with a monochrome detector. Compared with a quality, cooled, CCD detector, the loss may be more than 2 magnitudes, or around six times. Thus, whichever way you look at it, DSLRs do not go as deep as a cooled monochrome CCD.

However, to achieve any kind of color image that color data has to be acquired using filters somehow, and they always affect the image, making it very noisy. Fortunately, there is a solution, and it is called Luminance plus Red/Green/Blue (LRGB).

LRGB Imaging

In a nutshell, LRGB imaging gives you the best of both worlds when taking faint deep sky images. With a dedicated astrocamera, equipped with filters, you first take a totally unfiltered luminance image (or images), which reach as deep as possible. You then take red, green, and blue-filtered images that will look much noisier and reach less deep but, when combined together, using the LRGB command in popular packages such as, for example, Maxim DL (see Fig. 4.1), a miracle takes place. Those noisy filtered color images just melt way in the LRGB merger, and the final picture has the smooth, clean, and deep qualities of the unfiltered luminance image, combined with the color data from the filtered shots. Much of this miracle is down to the human eye/brain combination, which tends to see images in terms of brightness (luminance) first, with color pasted on top; thus, the noise in those color frames goes largely unnoticed. Indeed, such is the relative unimportance of the color image resolution

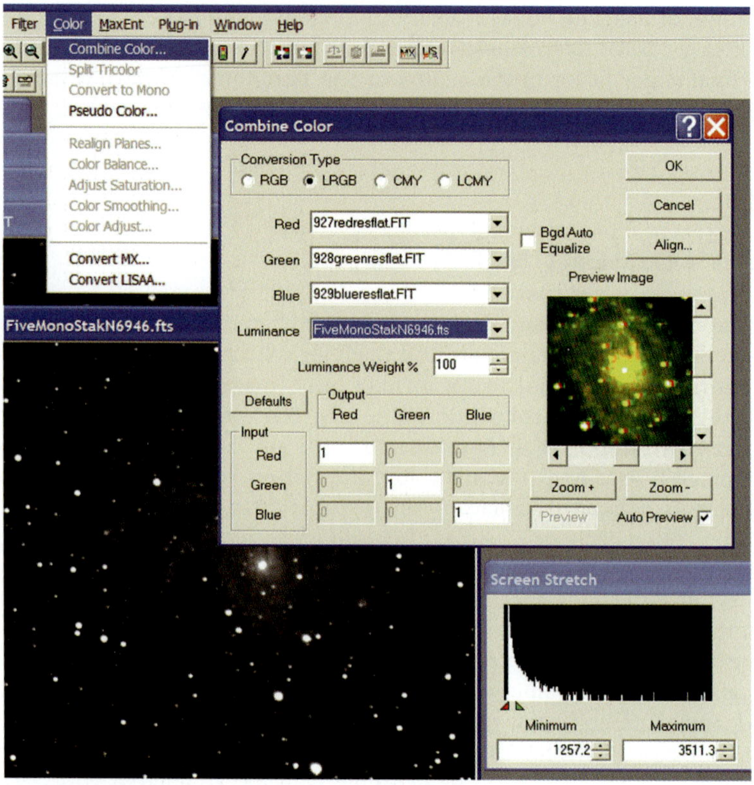

Fig. 4.1. The LRGB combination tool in the Maxim DL software package allows high quality color images to be assembled from a deep high contrast luminance image and noisier filtered color frames with relative ease.

that it is common for imagers to take their color shots binned 2×2, in other words, at half resolution and with the light from four pixels combined to get better signal-to-noise. Although the much smaller individual images look tiny compared with the full resolution luminance image, when combined using the LRGB method the lower color resolution goes unnoticed.

In recent years, even more LRGB and RGB tricks have been added to the imager's itinerary, especially where imaging emission line nebulae with narrow-band filters is concerned; see the section on imaging Caldwell emission nebulae for more details.

Imaging the Sky and Nothing Else!

We need to come clean at this point. We got a bit ahead of ourselves in describing how to take filtered and LRGB images, but there was a reason. Processing those images taken on great nights is far more exciting than the essential preprocessing chores generally known as dark frames and flat fields. We did not want to put the imager off. However, these topics do have to be covered.

But first, let us say a few words about archiving your raw images.

After a good imaging session every astronomer should treat his or her raw images like gold dust. Typically, these files are in FITS format or, for DSLR users, in the camera's own "RAW" format. However, they may also be in high quality, low compression JPEG format. Raw files are much larger than the highly compressed JPEG images that we are all familiar with. With imagers stacking a hundred or more images to make one deep picture they can have 100 megabytes of raw data just for one galaxy image. For the DSLR user an entire 1 gigabyte memory card can be eaten up in an evening! Nevertheless, the temptation to delete the raw files should be resisted when the images are great ones. CDs, DVDs, memory sticks, and portable hard drives are cheap these days, and those really clear nights are pretty rare events. In addition, the FITS image headers contain lots of useful exposure data that you will want to keep. With this in mind, let us get down to business...

Subtracting the Dark Frames

The raw images produced by CCD cameras can look very noisy and just plain ugly, as Fig. 4.2 proves! CCD cameras work by changing photons into electrons. The CCD camera's inbuilt "analog to digital converter" measures the number of electrons in each pixel, and this number is used to assign brightness values to each pixel on your PC screen.

Unfortunately, electrons are also generated simply by the CCD camera being warm. This thermal current "noise" doubles for every 7°C increase in temperature. Fortunately, as most commercial astro-cameras can be chilled by some 30°C, this cooling gives a 20-fold decrease in noise level. At an ambient temperature of 20°C many astronomical CCDs will saturate (white out) purely from thermal noise in a

Fig. 4.2. A raw image of NGC 891 (Caldwell 23) taken by the author before dark frame subtraction or flat field division.

10-min exposure. For DSLRs the CCD chip cannot be cooled, so you may be restricted to modest 10-min exposures on those mild summer nights, although, for many, genuine light pollution and the telescope drive would enforce this time limit anyway.

A dark frame is *the* solution to thermal noise; it is an exposure with the camera's shutter closed, and it captures the camera noise without the image of the astronomical object (and light pollution) being present. So, it might be thought that the temperature the camera was operating at was academic; surely you can just subtract it all using the dark frame, taken at the same temperature? Unfortunately, noise, by its very nature, is random. It is not a smooth level but a choppy sea. Subtracting a dark frame with the astro-camera's CCD running as cold as possible is definitely the best approach.

Dark frames (see Fig. 4.3) should be taken as close to the temperature at which the main image was exposed as possible; in practice this means immediately before, or after, the main image. Some astro-software packages arrange things so that a brand new automatic dark frame is exposed and subtracted if the exposure time has changed, or if the temperature changes by a degree or more, to ensure the best match subtraction. A few advanced amateurs build up a library of dark frames at various temperatures and exposures, along with a so-called "bias frame" (an exposure of almost zero duration, which records just the basic electronic read-out noise). Using this complex technique, it is possible to create custom dark frames to avoid very lengthy time-wasting exposures on those rare crystal-clear nights. For most situations though, simply taking one dark frame (once the camera and air temperature have stabilized) and sticking to that exposure duration all night, will be the best practical solution.

Not taking a dark frame is a worse crime than not cooling the camera. Although some of the lowest noise Sony CCDs and the Canon DSLR CMOS detectors can produce acceptable images without a dark frame subtraction, the ultra-sensitive Kodak

Fig. 4.3. A dark frame taken with the CCD camera shutter closed on the author's SBIG ST9XE CCD camera.

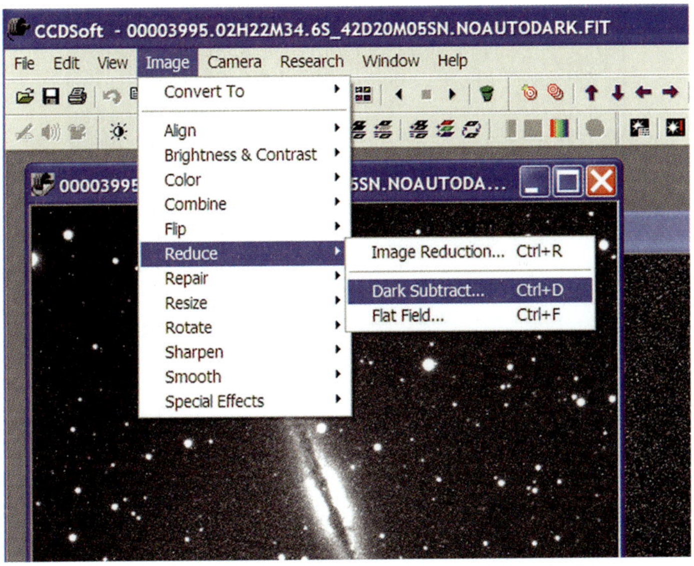

Fig. 4.4. A screen capture of Software Bisque's CCDSoft image reduction menu, used for dark subtraction or flat field division of raw astronomical format images.

"KAF" series CCDs used in many astro-cameras have considerable pixel-to-pixel variation; an image without a dark frame may be unusable, whereas an image at ambient temperature *with* a dark frame will produce a decent, if grainy, image. The built-in firmware for many DSLRs incorporates a "noise reduction mode" where a dark frame exposure is taken after the main image. Again, for identical exposures at the same temperature, only one dark frame is required per night; it can be reused on each frame. The software packages *IRIS*, *BlackFrame NR*, or Richard Berry's *AIP4Win* have useful dark frame subtraction routines for digital SLRs. Mike Unsold's *ImagesPlus* software is also popular in this regard. But you probably already have good dark frame routines with the camera or DSLR's firmware/software package. Figure 4.4 shows the CCDSoft menu structure for image reduction.

Dividing by the Flat Field

Flat fields are intended to make your final image background look a flat gray, because parts of the telescope light path can restrict (vignette) the light cone, especially when tele-compressors are involved. Also, dust specks (on glass surfaces above the chip) can appear as blurry doughnut-like shadows.

All good CCD astronomy packages incorporate an "Apply Flat Field" function that divides every pixel in the main image by every corresponding pixel in the flat field. The flat field image itself can be produced using smooth twilight, or by using a custom-built "light box" that fits over the telescope aperture. These boxes use low wattage bulbs and translucent perspex to produce an even background illumination. There are various Web sites about making flat field boxes, and Richard Berry's *AIP4Win* book also shows how to construct one.

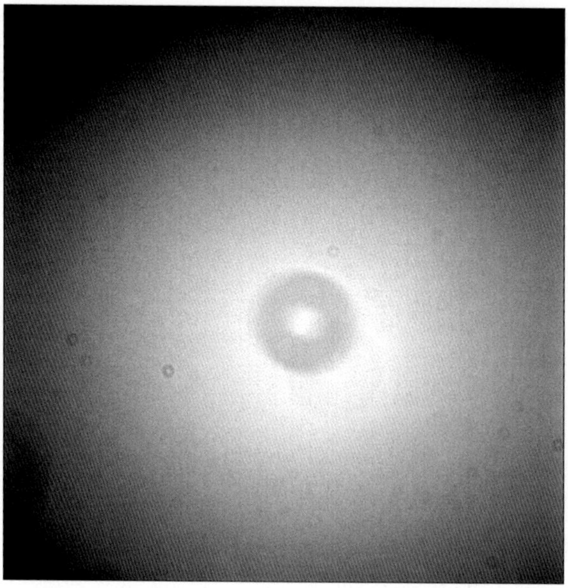

Fig. 4.5. A stack of multiple flat field images taken by the author by using subsecond exposures in a bright twilight sky. The light cut-off (vignetting) at the field edge is obvious as are the small and large doughnuts caused by specks of dust in the optics near the camera.

The twilight approach is very straightforward. First, you need a crystal clear postsunset sky. When the Sun is about 5° below the horizon, you can take very short exposures near the zenith, with your telescope, without the camera whiting out. The shorter the exposure the better, but some astronomy cameras with shutters do not let you go much below 0.2 s. You should aim to half-saturate the CCD to get a good signal-to-noise ratio. The exposures need to be short because stars can be captured, even in twilight, with longer exposures. The images obtained (see Fig. 4.5) will typically be brightest in the middle and covered with large and small rings, caused by dust specks. For a really excellent master flat field you should rattle off a few dozen flat (minus dark) frames and stack and average them. The resulting master flat field composite will be nice and smooth.

As well as applying these techniques it is worth checking out, in daylight or twilight but with the telescope aperture capped, whether your system is fully light-tight. Some uneven background field problems are actually caused by stray light (from streetlights, house lights, or observatory lights) entering the telescope or the CCD camera body at night. Once dark frame subtraction and flat field division has taken place, clean and flat images can be produced, such as in Fig. 4.6.

Autoguiding

Autoguiders are CCD cameras that, as well as imaging, can interface to a telescope drive and thereby keep a selected star from drifting across the camera pixels. In some SBIG cameras, an autoguiding CCD is mounted next to the imaging CCD. Why are they needed at all? Well, with the exceptions of the Paramount ME and the

Fig. 4.6. Each component of this image stack of NGC 891 (see also Fig. 4.2) has had the dark frame in Fig. 4.3 subtracted and then been divided by the flat field in Fig. 4.5, prior to stacking, to achieve this much cleaner and flatter result. This is the same image (taken by the author) that is used in Chap. 2 to illustrate the Caldwell 23 section.

largest Astrophysics mounts any amateur telescope that is imaging at finer than 2 arcsec per pixel will fail to keep stars on the same pixels for more than a minute or so. With a quality stainless steel worm and a big (>200 mm) phosphor bronze or aluminium wheel, exposures of two or more minutes start to become feasible, but with the smaller (70–140 mm diameter), lower quality wheels in mass-produced telescope mounts, a 30 or 60-s exposure is often the maximum limit, even with any periodic error correction (PEC) firmware engaged. Imaging objects at high declinations, say, above 60°, will halve the tracking errors.

It might be thought that an autoguider was the ultimate solution. Surely you just ask it to track a guide star at the edge of the main field and sit back while it controls the telescope? If only life were that simple! In practice most experienced amateurs simply determine the practical exposure limits of their equipment and then rattle off dozens of shots of that duration over a period of 10, 20, or 30 min. They then discard the trailed shots and stack the best ones using software such as *IRIS* or *Registax*. This approach certainly does not go as deep, but there is no autoguider set up time, either. With an autoguider, if you are imaging an object at, say, *f*/3.5 in the Milky Way, then you will be well served for guide stars. But at *f*/10 in star-barren Pegasus you may not find any suitable guide stars at all! Indeed, you may have to move the telescope slightly to bag a guide star. A lot of hassle!

Not all telescope mount manuals tell you precisely how to interface your autoguider either. Frustratingly, they say "autoguider compatible" in the advertising hype, but no interface or wiring advice comes with the manual. Helpful – NOT! Joining one of the online *Yahoo* telescope user groups is usually the best strategy in such situations. Even when full instructions are supplied, the degree to which the autoguider actually moves the telescope has to be customized in terms of speed and acceleration, so the drive reacts quickly, but does not simply overshoot in trying to keep the star centered. It can sometimes take 20 min to set an autoguider up, by which time cloud may have rolled in. Cloud is not a disaster if you are taking loads of short exposures. You just bin the cloud-polluted exposures and stack the rest of the best. But with one single long-autoguided exposure, a single cloud trashes the entire image.

In a nutshell, if you crave taking a few stunning pictures a year from a very dark site then using an autoguider on long, clear nights is the ultimate solution. However, if you just want to get reasonable deep sky or comet images every clear night, from a typical, grotty site, then taking hods of 30 or 60 s shots, and stacking the sharpest will be the simplest practical solution. As mentioned much earlier a single exposure of, say, 2 min will go much deeper than, say, eight individual 15-s exposures stacked together. Nevertheless, stacking images does have a significant advantage in reducing noise and smoothing the picture, as much of the noise in each image is random.

An additional step can be taken, too, and this is known as dithering. Slewing the telescope by just a few pixels, randomly, between images can get rid of the obstinate fixed pattern noise that tends to be obvious even if the telescope tracking drifts steadily between exposures. By stacking randomly micro-slewed frames on top of each other a very smooth image can result. Obviously, if an autoguider is used this dithering process can be problematic, as the autoguider may need to adjust to a new guide star position between exposures. Images can be stacked in various ways, but a simple averaging process is usually preferred over an addition process so that bright nebulae do not end up being saturated (whited out). It goes without saying that when "dithering" lots of images the total dither should not cause the field to drift in the same direction by scores of pixels, as this will noticeably reduce the available field of view.

Now let us look at some specific imaging techniques applied to different categories of Caldwell objects.

Imaging the Caldwell Open Clusters

Star clusters, both open and globular, are the easiest deep sky objects to image and open clusters are the easiest of all. They consist of relatively bright stars occupying relatively large areas of sky and so can be captured in small or modest telescopes with no more equipment than a digital SLR. However, there is quite a range of cluster diameters to cope with, and so no one instrument is going to suffice for all. The smallest of the 28 Caldwell open clusters are just 6 or 7 arcmin across, but the largest object, the Hyades cluster (Caldwell 41), spans at least 5° and so is best imaged with a telephoto lens, not a telescope. For the most part, an image of an open cluster will look just as good in monochrome as in color. Admittedly, a few of the Caldwell open clusters feature highly colored stars within the swarm, but most do not, and it is all too easy to saturate the stars, making them all look white anyway.

For artistic effect, some astrophotographers prefer to image open clusters using a small Newtonian with a 4-vane secondary mirror holder such that bright stars show the "traditional" cross-shaped spikes caused by diffraction; however, such special effects can quite easily be added with image processing software trickery!

In addition, users of scientific grade CCDs without the commercial "antiblooming gates" on the chip may be plagued with the "line bleeding" caused by such detectors when they saturate.

Some wide angle starfield astrophotography specialists have found that placing an appropriate photographic diffusing filter in front of a telephoto lens bloats the diameter of the brightest stars considerably, making them appear as significant as they do to the human eye and, because the intensity is reduced, desaturating the color from an overexposed white to the true color of the star. Such tricks can lead to truly beautiful wide angle shots, such as produced by the Japanese astrophotographer Akira Fujii. It is, perhaps, worth mentioning the subject of CCD antiblooming gates in this section on imaging open clusters. Open clusters invariably contain a few very bright stars. When a CCD sensor incorporates antiblooming gates (ABG) on its surface any excess electrical charge from bright stars that saturate the detector's pixels will be drained off as soon as the pixels approach saturation. Without antiblooming gates (NABG) the charge overflows the pixel and bleeds a white line down the column on the chip. On this basis it might seem a good idea to buy an ABG type detector, as it will lead to a pretty picture (as with a digital SLR). However, you need to bear in mind that NABG devices become unreliable for use in scientific photometric applications when the pixels exceed half-saturation. The alternate approach would be to expose images with short enough exposures that not even the brighest stars bleed; then simply stack and average all the short exposures.

Imaging the Caldwell Globular Clusters

Only 18 of the 109 objects in the Caldwell catalog are globular clusters, and the vast majority are in the southern hemisphere; this is hardly surprising, as Messier spotted the best 29 globulars visible from Paris, and the two catalogs are mutually exclusive. Some of the finest southern hemisphere globular images in this book were taken by Daniel Verschatse (see Fig. 4.7)

The Caldwell globulars cover a huge range of magnitudes and angular sizes from 10.6 to 3.7 magnitude and from 3.6 to 55 arcmin, if we take the extreme cases of NGC 7006 (C42) and Omega Centauri (NGC 5139/C80). Perhaps more relevant are the typical star magnitudes encountered; the brightest range from 11th to 15th in the nearest and furthest examples, respectively, and in those furthest cases the stars are hard to split even with big amateur telescopes, great guiding, and good seeing. Patrick's choice of objects has created something of a glut of globulars at certain southerly declinations, with eight examples being available from C73 to C87 and all five objects from C104 to C108 are globulars, too, in the southern Polar Regions.

Of all the deep sky objects easily visible in amateur telescopes, the globular clusters are, arguably, the most mesmerizing. They are spherical aggregations of mainly ancient stars that are gravitationally bound to our Milky Way but live outside the main plane of the galaxy, in the halo, in fact. Globular clusters live thousands or tens of thousands of light-years from our Solar System, and the largest examples may contain more than a million stars and span more than 100 light-years in size.

Fig. 4.7. Many of the best globular cluster and other southern hemisphere deep sky images in this book were supplied by Daniel Verschatse who images from a dark site at Antilhue in Chile. Daniel is shown here with his largest instrument, a 14.5-in. (37 cm) RCOS Ritchey-Chrétien and an Astrophysics AP 130EDF refractor mounted on an Astrophysics AP1200-GTO mount. © Daniel Verschatse – Observatorio Antilhue – Chile.

In some ways these celestial jewel boxes can be thought of as companion mini-galaxies, but morphologically they are totally different from, for example, the more distant Large and Small Magellanic clouds, visible in the southern hemisphere. Globulars are all perfectly spherical in appearance, with the concentration of stars becoming increasingly dense toward the center, at least in most cases. Visually, the challenge is seeing whether you can detect the subtle stellar patterns described by keen deep sky observers in the popular literature, and whether individual stars can be observed, too. Remarkably, there are now a total of 158 globular clusters associated with our Milky Way, although only a third of these can be classed as easy amateur targets.

The appearance of a globular can be categorized using the system of Shapley and Sawyer. This classification uses Roman numerals such that types I to III have an obvious very high stellar density at their cores, types IV to VI have a lesser core density, VII to IX just have a subtly denser core, and types X to XII look, essentially, evenly dense across the globular's diameter. Globulars contain some of the oldest (and lowest metal content) stars in the environment of our galaxy. Living in their cores alien deep sky observers would have a very tough time! When the Hubble

Space Telescope peered into the center of the massive globular Omega Centauri astronomers deduced there were some 5,000 stars visible in a region only 13 light-years across!

As we have seen, on their respective Caldwell pages, Omega Centauri (NGC 5139/Caldwell 80) and 47 Tucanae (NGC 104/Caldwell 106) are truly magnificent and make the northern hemisphere's M13 look rather feeble by comparison. At magnitude 3.7 Omega Centauri is a naked-eye object containing millions of stars, and it is perfectly placed in southern hemisphere May skies. At a declination of minus 47° you do not have to cross the equator to spot it. Even from Tenerife it can rise 14° above the southern horizon. You have to travel much further south to see 47 Tucanae, though, as it has a declination of −72° and is best placed in September, not May. It lies just to the west of the Small Magellanic Cloud (SMC), a genuine companion galaxy of the Milky Way. At a distance of 13,400 light-years and containing several million stars, the glorious 47 Tucanae is 15× nearer than the SMC but has only a hundredth the number of stars.

Imaging Strategy

Unlike nebulae and supernova remnants, dramatic images of globular clusters can be secured without resorting to narrowband filters and cumulative exposures of hours. They are, after all, broadband objects (as stars emit light over the whole visual spectrum), and they have a healthy surface brightness as well as a healthy diameter of many arcminutes (almost a degree in the case of Omega Centauri). In addition, because it is not helpful to saturate the core, it is possible to overexpose a globular cluster image. The individual stars in all of the Caldwell globulars are intense points, and even when they are 16th or 17th magnitude a big amateur telescope can image them clearly in under 30 s with a cooled CCD camera.

Perhaps the best news is that globular clusters look just as good in monochrome as in color, making imaging them even easier. However, to get the best images some effort is always necessary, and with globulars this effort should be concentrated on the guiding and tracking, as pinpoint star images are vital. With the brightest globular clusters, it is all too easy to saturate the core, so it is far better to take multiple short exposures and average them in this situation, something which will be a great relief to those with mediocre telescope drives.

Because resolution is all important in the imaging of globulars a long focal length will deliver the best results. However, Omega Centauri (C 80) and 47 Tucanae (C 106) are so large (55′ and 50′) that a long focal length with a small CCD detector may cause the globular to overflow the field. The other Caldwell globulars are much smaller and, in practice, the official diameters of globulars are rather bigger than they appear visually or even on amateur images. In some ways, monochrome images of globular clusters look even more dramatic than color shots. To get a punchy image, but not one where the core is too saturated, you will want to play about with the contrast, brightness, gamma, and log scaling functions in your image processing package. All in all, if you have a long focal length, good tracking, and sharp focusing then the Caldwell globulars will never disappoint and, compared with more nebulous objects, are easy to capture well.

Imaging the Caldwell Emission and Planetary Nebulae

Deep sky objects that are nebular in nature (in other words, excluding open clusters, globular clusters, and galaxies) tend to emit significant quantities of light in the narrow astronomical emission bands that occupy a tiny part of the spectrum. Therefore, especially when light pollution is taken into account, it can pay big dividends to just image in these narrow wavelengths with the available commercial filters, to get a huge signal-to-noise gain.

Great results using narrowband filters can be obtained with relatively modest apertures, such as those used by the deep sky imager Ian Sharp in his observatory; see Fig. 4.8. Good narrowband H-alpha (Hydrogen) images can even be taken in moonlight, as only a small fraction of the scattered light in our atmosphere pollutes the narrow H-alpha bandwidth. It should perhaps be stressed here that while deep sky H-alpha filters are narrowband (typically 3–7 nm), they are nowhere near as narrow as the H-alpha filters used for solar work, which span a

Fig. 4.8. Some fine emission nebula images in this book were obtained by Ian Sharp of Ham, West Sussex, who lives just a short distance from the Caldwell list inventor, Sir Patrick Moore. This is Ian's wide-field set up: an Astrophysics AP900-GTO mount with an Orion ED80 semi-apo refractor and an 80 mm SkyWatcher guide scope, mounted inside Ian's homemade 3.2 m observatory dome (wooden walls and floor with a fiberglass dome).

hundredth of the bandwidth of the deep sky filters (being less than one Angstrom bandwidth).

Of course, by everyday standards, all H-alpha nebulosity is painfully feeble, and using a filter does not get around this; it just darkens the sky background enormously, thus increasing the contrast dramatically. A single H-alpha image of a suitable nebula or supernova remnant often consisting of many hours of cumulative exposures is only a monochrome image, and a color image makes a much prettier picture. For this reason amateurs have, with much success, used H-alpha luminance images, combined with red, green, and blue images to produce a colorful LRGB picture. In most cases, the red contribution has also used the H-alpha data, with the green and blue images being provided by normal color filters or, sometimes, with the green being synthetically produced from an average of the H-alpha and blue data!

The H-alpha region is right at the far end of the human eye's red response, and so these filters are not intended for visual use with such ghostly objects. However, CCDs have a very healthy response at these wavelengths, and so H-alpha filters, when used with a cooled monochrome detector, work very well. With some nebulae that have a strong OIII (Oxygen) emission an OIII filtered image has sometimes been used to provide the green data in LRGB images. Some amateurs use SII (Sulfur) band filters, too, an even deeper red emission band than H-alpha, which is sometimes appropriate when imaging supernova remnants or planetary nebulae. The respective wavelengths of these filters are 6,563 Å (656.3 nm) for H-alpha, 5,007/4,959 Å (500.7 and 495.9 nm) for OIII, and 6,719/6,730 Å (671.9/673.0 nm) for SII.

Beyond LRGB

Unfortunately, even the powerful LRGB technique is not perfect, primarily because of the effect it has on stars, especially when a narrowband filter is used for luminance data. Emission nebulae are powerfully enhanced with respect to the sky background in H-alpha images, but the stars are heavily attenuated, as they are broadband emitters; thus, star sizes and colors become badly distorted when H-alpha is used for the luminance data. So some amateurs now prefer to use a standard RGB technique but one in which the R component is partly real R data and partly H-alpha data. For example, a red image averaged with a deep H-alpha image would produce a much more natural, if less contrasty, RGB result. The same argument applies to OIII data, where one could use a 50:50 OIII/green combination. Indeed, virtually any combination is possible, and it is true that the amateurs who get the best results never cease experimenting with every image they take. But, in each case, the essential requirement is to get some deep, sharp, and clean master images using all the filters you plan to use on the chosen object.

The narrowband LRGB advice in the previous section applies to planetary nebulae as well as general emission nebulae, but there is one extra factor where this subcategory of nebulae is concerned: planetary nebulae are usually *very* small. Admittedly their diminutive size also means they have a very high surface brightness, too, which is great for the visual observer. However, they are so small that critical focusing and tracking are essential to get any interior detail.

In addition, atmospheric seeing plays a part, too. Without access to the best equipment one strategy is simply to take scores of short (say, 5-s) exposures and

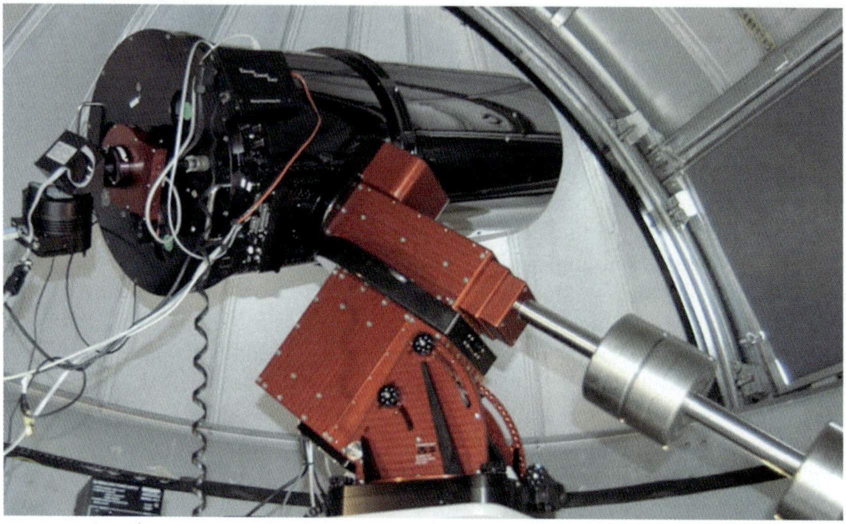

Fig. 4.9. Gordon Rogers of Long Crendon, UK supplied a number of splendid images for this book. Some were taken with his original 0.4 m Schmidt-Cassegrain, but the more recent ones were captured with his magnificent 0.4 m RCOS Ritchey-Chrétien pictured here. The instrument is mounted in a 10′ 6″ (3.2 m) Ash dome built on top of Gordon's house. http://www.gordonrogers.co.uk

stack and average them. With dim deep sky objects such exposures would be very noisy, but there is more than enough signal-to-noise ratio to use such short exposures on the high surface brightness planetary nebula. An additional benefit is that saturating the bright nebular detail is avoided with this technique. With access to the best equatorial mounts, exemplified by the Paramount ME, or adaptive optics units, such as SBIG's AO-7 and AO-8, long autoguided exposures through narrowband filters at very long focal lengths can be used with these tiny planetary nebulae. In many cases, the resulting image will still look small, but pixel resampling by, say, 300% can easily cure this. One example of a top notch piece of kit for long focal length imaging is shown in Fig. 4.9.

Imaging and Supernova Patrolling the Caldwell Galaxies

LRGB, LLRGB, and DDP

We have already seen that the LRGB technique works well by combining the raw unfiltered penetration of a monochrome image with the noisier colored data from the filtered images. Galaxies benefit well from this technique, and as they are not narrowband emitters standard R, G, and B color filters can be used to good effect. You can delay the expenditure on those pricey H-alpha, OIII, and SII filters if galaxies are your favorite targets.

In recent years, some amateurs, most notably Robert Gendler in the United States – arguably the world's greatest deep sky imager – have taken the LRGB technique one stage further on galaxies and produced an LLRGB method. This technique produces a mild LRGB image of an object first, with the luminance contribution down to about 50% and the extremes of brightness reduced while the color saturation is boosted. The R, G, B values from this resulting image are then used as the R, G, B values in a full blown LRGB image. If all this sounds very complicated, do not panic! In reality, every image displayed on a PC screen only consists of red, green, and blue stripes; how brightly each of these stripes glows is simply a function of the formula in the software used to produce LRGB images. That function may exist, as in astro-software, as an LRGB box or, as in *Photoshop*, as a luminosity/opacity option in the layers menu.

Another feature that may well appeal to the galaxy imager is Digital Development Processing or DDP. A "DDP tool" may well be present in your preferred software package. The problem with really awesome galaxies is that they have much brighter centers than outer spiral arms. Boost the contrast, and the spiral arms come out nicely, but the core bleaches out to a solid white, almost to the inner start of the spiral arms in some cases.

As there is often some superb detail in the central regions of galaxies, this can all be lost in a simple contrast stretch. What is needed is a stretch that brightens the faint bits but does not brighten or contrast stretch the inner galaxy core as much. Photographic film used to have a nonlinear response to light (a nonlinear "gamma curve") that worked well in this regard, as well as an edge-sharpening response that worked particularly well on galaxies. The DDP algorithm effectively adjusts the relationship between brightness levels in the raw image (typically 16 bit or 65,536 gray levels) and the displayed image on your PC (typically 8 bit or 256 gray levels), so the contrast of faint features is enhanced but not at the expense of the brighter parts. In addition, DDP also throws in an edge sharpening effect similar to an unsharp mask, which sharpens up fine details and makes those stars look pinsharp, rather than bloated. DDP is not the only solution for galaxy image processing, though, and most astronomy software packages feature other contrast stretching algorithms, such as a gammalog stretch, which can produce a similar effect. The AIP4Win menu structure for DDP is shown in Fig. 4.10.

Supernova Hunting

Of course when amateur astronomers think about galaxies they automatically think about supernova hunting. I have covered this aspect of the hobby fully in my other book in this series entitled 'Supernovae and How to Observe Them' but it is worth revisiting the prospects for success with the Caldwell galaxies in this section. There are 35 galaxies in the Caldwell catalog and 16 of them (see Table 4.1) have produced supernovae during the last century – a rather poorer ratio than with the Messier galaxies, as would be expected from a sample containing more neglected southern hemisphere objects.

Nevertheless, single-handedly Caldwell 12 (NGC 6946) makes up for this with its ridiculous total of nine supernovae, the latest one (SN 2008 S) being discovered by UK patroller Ron Arbour.

Apart from Caldwell 12/NGC 6946, the other most productive Caldwell galaxies are C7, C60, and C67, namely NGC 2403, NGC 4038, and NGC 1097, respectively. One of the brightest supernovae of all time was discovered in NGC 2403 as recently as July 31, 2004, by the Japanese amateur astronomer Itagaki. SN 2004dj

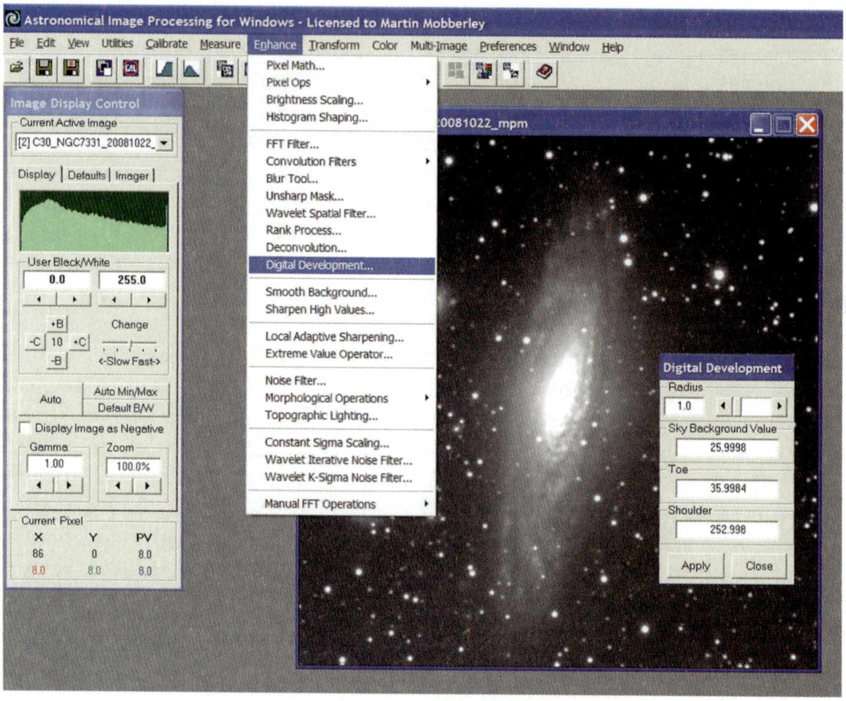

Fig. 4.10. The DDP (Digital Development) tool under AIP4Win's "Enhance" menu is useful for bringing out details in faint spiral arms, suppressing bright galaxy cores and sharpening star images.

Table 4.1. Caldwell galaxy supernovae up to 2008

Galaxy	Supernovae	Amateur Caldwell discoverers
C7	2004dj; 2002 kg; 1954 J (super outburst)	Itagaki (04dj)
C12[a]	2008 S; 2004et; 2002hh; 1980 K; 1969P; 1968D; 1948B; 1939C; 1917A	Arbour (08 S) Moretti(04et)
C23	1986 J	
C24	2005mz; 1968A	Newton et al.
C29	1996ai	
C30	1959D	
C36	1941A	
C44	1990U	
C48	1993Z	
C53	1935B	
C60	2007 sr; 2004gt; 1974E; 1921A	Monard (04gt)
C65	1940E	
C67	2003B; 1999eu; 1992bd	Evans (03B); Aoki (99eu)
C77	1986G	Centaurus A SN by Evans
C83	2005af	
C101	2005at	Monard co-discovery

[a] C12, alias NGC 6946, is, of course, the most SN productive galaxy in the entire sky, with nine discoveries from 1917 to 2008 in various parts of that galaxy

was a considerable distance (160″ east and 10″ north) from the galaxy's nucleus, meaning that its decline from maximum brightness was well observed.

Remarkably, only 2 years earlier, another supernova was discovered in NGC 2403, the much fainter SN 2002 kg. Being situated at 65° north means that this galaxy is circumpolar, and therefore it can be patrolled all year round from northern Europe and the northern United States/Canada. Supernova 1954 J in Caldwell 7/NGC 2403 has, in recent years, proved not to be a supernova at all. The star, also known as "Variable 12" in that galaxy, actually survived what appears to have been a superoutburst, similar to the 1843 eruption of eta Carinae in our Milky Way. The apparent 25 solar-mass survivor can still be imaged with the Hubble Space Telescope and has an absolute magnitude now of about –8.

Caldwell 60, which produced supernovae 2007 sr, 2004gt, 1974E, and 1921A is actually just the northwestern component of the superb southern hemisphere spectacle called the Antennae (those two galaxies colliding with each other). As we have seen the other, southeastern, component is designated as Caldwell 61. Their NGC numbers are 4038 and 4039. At declinations of almost –19°, the Antennae can be seen from latitudes as high as the UK but only as they transit in a spring sky and, even then, never more than 20° above the horizon. Caldwell 67, the only other Caldwell galaxy with at least three supernovae, is another southern hemisphere galaxy, and all of its supernovae were discovered in an 11 year span from 1992 to 2003, two by amateur hunters Evans and Aoki.

Caldwell Resources

Caldwell Books

Deep-Sky Companions – The Caldwell Objects, by Stephen James O'Meara, published by Sky & Telescope/Cambridge University Press in 2002.

Observing the Caldwell Objects, by David Ratledge, published by Springer in 2000.

Caldwell Websites

Wikipedia Caldwell page
http://en.wikipedia.org/wiki/Caldwell_catalog

Students for the Exploration and Development of Space Caldwell page
http://ftp.seds.org/Messier/xtra/similar/caldwell.html

Utah Skies page
http://www.utahskies.org/deepsky/caldwell/

Hawaiian Astronomical Society Caldwell page
http://www.hawastsoc.org/deepsky/caldwell.html

David Richards' Caldwell page
http://www.richweb.f9.co.uk/astro/caldwell_objects.htm

David Paul Green's Caldwell Object Log software
http://www.davidpaulgreen.com/tcol.html

Images and Telrad Maps for Caldwell Objects
http://www.astro-tom.com/advanced/caldwell/maps_to_caldwell_objects.htm

Space and Telescope Caldwell page
http://www.space-and-telescope.com/CaldwellObjects.aspx

Planetarium Software for Creating Custom Charts

Project Pluto's Guide 8.0
http://www.projectpluto.com/

Chris Marriott's SkyMap
http://www.skymap.com/

Software Bisque's The Sky
http://www.bisque.com/Products/TheSky6/

Starry Night Software
http://www.starrynightstore.com/

Deep Sky Imagers' Websites

Adam Block – Caelum Observatory
http://www.caelumobservatory.com/

Kitt Peak Advanced Observing Program – Best images
http://www.noao.edu/outreach/aop/bestof.html

Capella Observatory – Dr. Stefan Binnewies and Josef Pöpsel
http://www.capella-observatory.com/

Tom Carrico
http://www.ccdargo.com/

Sven Kohle and Till Credner
http://www.allthesky.com/

Christian Fuchs, Philipp Keller and the T1T-Team
http://www.astrooptik.com/
and
http://www.astrooptik.com/Bildergalerie/PolluxGallery/PolluxGallery.htm

Robert Gendler
http://www.robgendlerastropics.com/

Tom Harrison
http://www.manoprietoobservatory.com/

Gordon Rogers
http://www.gordonrogers.co.uk/

Ian Sharp
http://www.astro-sharp.com/

Daniel Verschatse
http://www.verschatse.cl/

Volker Wendel and Bernd Flach-Wilken of the Spiegel Team
http://www.spiegelteam.de/

Geof Wingham
http://www.pbase.com/loc46south/profile

Deep Sky Visual Observers' Websites

Faith Jordan's site
http://visualdeepsky.webs.com/index.htm

Iiro Sairanen's site
http://personal.inet.fi/surf/deepsky/

Barbara Wilson's site
http://home.ix.netcom.com/~bwilson2/barbarasweb/

Visual Observing Equipment

TeleVue eyepieces and quality refractors
http://www.televue.com/engine/page.asp?ID=2

Obsession Dobsonians
http://www.obsessiontelescopes.com/

Starmaster Dobsonians
http://www.starmastertelescopes.com/

Orion Optics (UK) Dobsonians
http://www.orionoptics.co.uk/DOBSONIAN/dobsonianrange.html

Visual and imaging deep sky filters distributed by OPT
https://www.optcorp.com/category.aspx?uid=105-109-156-823

Image Processing Astro-software

IRIS (powerful freeware)
http://www.astrosurf.com/buil/us/iris/iris.htm

AIP4Win (book and software)
http://www.willbell.com/aip/index.htm

CCDSoft
http://www.bisque.com/help/ccdsoft%20info/CCDSoft_Information_Site.htm

MaximDL
http://www.cyanogen.com/maxim_main.php

ImagesPlus
http://www.mlunsold.com/

Registax
http://www.astronomie.be/registax/

BlackFrame NR
http://www.mediachance.com/digicam/blackframe.htm

High Quality Deep Sky Astrographs and Mounts

RC Optical Systems
http://www.rcopticalsystems.com/

Optical Guidance Systems
http://www.opticalguidancesystems.com/

Planewave Instruments
http://www.planewaveinstruments.com/

Astrophysics' apochromats and mountings
http://www.astro-physics.com/

Takahashi
http://takahashiamerica.com/catalog/

Paramount ME mounting
http://www.bisque.com/Products/Paramount/

CCD Cameras

Santa Barbara Instruments Group (SBIG)
http://www.sbig.com/sbwhtmls/online.htm

SBIG Users Group
http://tech.groups.yahoo.com/group/SBIG/

Starlight Xpress
http://www.starlight-xpress.co.uk/

Apogee
http://www.ccd.com/

Atik
http://www.atik-instruments.com/

Index

Index

Printed in the United States of America